T0327644

Introduction to Random Signals and Noise

Introduction to Random Signals and Noise

Wim C. van Etten
University of Twente, The Netherlands

John Wiley & Sons, Ltd

Telephone (+44) 1243 779777

Email (for orders and customer service enquiries): cs-books@wiley.co.uk
Visit our Home Page on www.wiley.com

Reprinted with corrections March 2006

Other Wiley Editorial Offices

John Wiley & Sons Inc., 111 River Street, Hoboken, NJ 07030, USA

Jossey-Bass, 989 Market Street, San Francisco, CA 94103-1741, USA

Wiley-VCH Verlag GmbH, Boschstr. 12, D-69469 Weinheim, Germany

John Wiley & Sons Australia Ltd, 42 McDougall Street, Milton, Queensland 4064, Australia

John Wiley & Sons (Asia) Pte Ltd, 2 Clementi Loop #02-01, Jin Xing Distripark, Singapore 129809

John Wiley & Sons Canada Ltd, 22 Worcester Road, Etobicoke, Ontario, Canada M9W 1L1

Wiley also publishes its books in a variety of electronic formats. Some content that appears in print may not be available in electronic books.

British Library Cataloguing in Publication Data

A catalogue record for this book is available from the British Library

ISBN-13 978-0-470-02411-9 (HB)
ISBN-10 0-470-02411-9 (HB)

Typeset in 10/12pt Times by Thomson Press (India) Limited, New Delhi

To Kitty,
to Sascha, Anne and Emmy,
to Björn and Esther

Contents

Preface

Random signals and noise are present in several engineering systems. Practical signals seldom lend themselves to a nice mathematical deterministic description. It is partly a consequence of the chaos that is produced by nature. However, chaos can also be man-made, and one can even state that chaos is a *conditio sine qua non* to be able to transfer information. Signals that are not random in time but predictable contain no information, as was concluded by Shannon in his famous communication theory.

To deal with this randomness we have to nevertheless use a characterization in deterministic terms; i.e. we employ probability theory to determine characteristic descriptions such as mean, variance, correlation, etc. Whenever chaotic behaviour is time-dependent, as is often the case for random signals, the time parameter comes into the picture. This calls for an extension of probability theory, which is the theory of stochastic processes and random signals. With the involvement of time, the phenomenon of frequency also enters the picture. Consequently, random signal theory leans heavily on both probability and Fourier theories. Combining these subjects leads to a powerful tool for dealing with random signals and noise.

In practice, random signals may be encountered as a desired signal such as video or audio, or it may be an unwanted signal that is unintentionally added to a desired (information bearing) signal thereby disturbing the latter. One often calls this unwanted signal noise. Sometimes the undesired signal carries unwanted information and does not behave like noise in the classical sense. In such cases it is termed as interference. While it is usually difficult to distinguish (at least visually) between the desired signal and noise (or interference), by means of appropriate signal processing such a distinction can be made. For example, optimum receivers are able to enhance desired signals while suppressing noise and interference at the same time. In all cases a description of the signals is required in order to be able to analyse their impact on the performance of the system under consideration. In communication theory this situation often occurs. The random time-varying character of signals is usually difficult to describe, and this is also true for associated signal processing activities such as filtering. Nevertheless, there is a need to characterize these signals using a few deterministic parameters that allow a system user to assess system performance.

This book deals with stochastic processes and noise at an introductory level. Probability theory is assumed to be known. The same holds for mathematical background in differential and integral calculus, Fourier analysis and some basic knowledge of network and linear system theory. It introduces the subject in the form of theorems, properties and examples. Theorems and important properties are placed in frames, so that the student can easily summarize them. Examples are mostly taken from practical applications. Each chapter concludes with a summary and a set of problems that serves as practice material. The book is well suited for dealing with the subject at undergraduate level. A few subjects can be skipped if they do not fit into a certain curriculum. Besides, the book can also serve as a reference for the experienced engineer in his daily work.

In Chapter 1 the subject is introduced and the concept of a stochastic process is presented. Different types of processes are defined and elucidated by means of simple examples.

Chapter 2 gives the basic definitions of probability density functions and includes the time dependence of these functions. The approach is based on the 'ensemble' concept. Concepts such as stationarity, ergodicity, correlation functions and covariance functions are introduced. It is indicated how correlation functions can be measured. Physical interpretation of several stochastic concepts are discussed. Cyclo-stationary and Gaussian processes receive extra attention, as they are of practical importance and possess some interesting and convenient properties. Complex processes are defined analogously to complex variables. Finally, the different concepts are reconsidered for discrete-time processes.

In Chapter 3 a description of stochastic processes in the frequency domain is given. This results in the concept of power spectral density. The bandwidth of a stochastic process is defined. Such an important subject as modulation of stochastic processes is presented, as well as the synchronous demodulation. In order to be able to define and describe the spectrum of discrete-time processes, a sampling theorem for these processes is derived.

After the basic concepts and definitions treated in the first three chapters, Chapter 4 starts with applications. Filtering of stochastic processes is the main subject of this chapter. We confine ourselves to linear, time-invariant filtering and derive both the correlation functions and spectra of a two-port system. The concept of equivalent noise bandwidth has been defined in order to arrive at an even more simple description of noise filtering in the frequency domain. Next, the calculation of the spectrum of random data signals is presented. A brief resumé of the principles of discrete-time signals and systems is dealt with using the z-transform and discrete Fourier transform, based on which the filtering of discrete-time processes is described both in time and frequency domains.

Chapter 5 is devoted to bandpass processes. The description of bandpass signals and systems in terms of quadrature components is introduced. The probability density functions of envelope and phase are derived. The measurement of spectra and operation of the spectrum analyser is discussed. Finally, sampling and conversion to baseband of bandpass processes is discussed.

Thermal noise and its impact on systems is the subject of Chapter 6. After presenting the spectral densities we consider the role of thermal noise in passive networks. System noise is considered based on the thermal noise contribution of amplifiers, the noise figure and the influence of cascading of systems on noise performance.

Chapter 7 is devoted to detection and optimal filtering. The chapter starts by considering hypothesis testing, which is applied to the detection of a binary signal disturbed by white Gaussian noise. The matched filter emerges as the optimum filter for optimum detection performance. Finally, filters that minimize the mean squared error (Wiener filters) are derived. They can be used for smoothing stored data or portions of a random signal that arrived in the past. Filters that produce an optimal prediction of future signal values can also be designed.

Finally, Chapter 8 is of a more advanced nature. It presents the basics of random point processes, of which the Poisson process is the most well known. The characteristic function plays a crucial role in analysing these processes. Starting from that process several shot noise processes are introduced: the homogeneous Poisson process, the inhomogeneous Poisson process, the Poisson impulse process and the random-pulse process. Campbell's theorem is derived. A few application areas of random point processes are indicated.

The appendices contain a few subjects that are necessary for the main material. They are: signal space representation and definitions of attenuation, phase shift and decibels. The rest of the appendices comprises basic mathematical relations, a summary of probability theory, definitions of special functions, a list and properties of Fourier transform pairs, and a few mathematical and physical constants.

Finally, I would like to thank those people who contributed in one way or another to this text. My friend Rajan Srinivasan provided me with several suggestions to improve the content. Also, Arjan Meijerink carefully read the draft and made suggestions for improvement.

Last but certainly not least, I thank my wife Kitty, who allowed me to spend so many hours of our free time to write this text.

Solutions to problems set in the book can be found at: www.wiley.com/go/randomsignalsand noise.

Wim van Etten
Enschede, The Netherlands

1 Introduction

1.1 RANDOM SIGNALS AND NOISE

In (electrical) engineering one often encounters signals that do not have a precise mathematical description, since they develop as random functions of time. Sometimes this random development is caused by a single random variable, but often it is a consequence of many random variables. In other cases the causes of randomness are not clear and a description is not possible, but the signal is characterized by means of measurements only.

A random time function may be a desired signal, such as an audio or video signal, or it may be an unwanted signal that is unintentionally added to a desired (information) signal and disturbs the desired signal. We call the desired signal a random signal and the unwanted signal noise. However, the latter often does not behave like noise in the classical sense, but it is more like interference. Then it is an information bearing signal as well, but undesired. A desired signal and noise (or interference) can, in general, not be distinguished completely; by means of well-defined signal processing in a receiver, the desired signal may be favoured in a maximal way whereas the disturbance is suppressed as much as possible. In all cases a description of the signals is required in order to be able to analyse its impact on the performance of the system under consideration. Especially in communication theory this situation often occurs. The random character as a function of time makes the signals difficult to describe and the same holds for signal processing or filtering. Nevertheless, there is a need to characterize these signals by a few deterministic parameters that enable the system user to assess the performance of the system. The tool to deal with both random signals and noise is the concept of the stochastic process, which is introduced in Section 1.3.

This book gives an elementary introduction to the methods used to describe random signals and noise. For that purpose use is made of the laws of probability, which are extensively described in textbooks [1–5].

1.2 MODELLING

When studying and analysing random signals one is mainly committed to theory, which however, can be of good predictive value. Actually, the main activity in the field of random signals is modelling of processes and systems. Many scientists and engineers have

Introduction to Random Signals and Noise W. van Etten

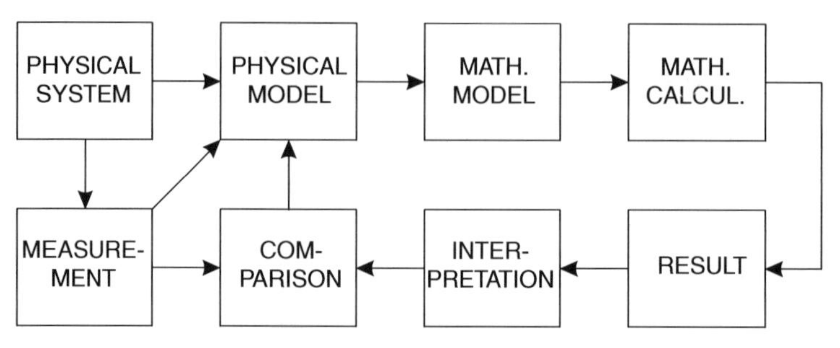

Figure 1.1 The process of modelling

contributed to that activity in the past and their results have been checked in practice. When a certain result agrees (at least to a larger extent) with practical measurements, then there is confidence in and acceptance of the result for practical application. This process of modelling has schematically been depicted in Figure 1.1.

In the upper left box of this scheme there is the important physical process. Based on our knowledge of the physics of this process we make a physical model of it. This physical model is converted into a mathematical model. Both modelling activities are typical engineer tasks. In this mathematical model the physics is no longer formally recognized, but the laws of physics will be included with their mathematical description. Once the mathematical model has been completed and the questions are clear we can forget about the physics for the time being and concentrate on doing the mathematical calculations, which may help us to find the answers to our questions. In this phase the mathematicians can help the engineer a lot. Let us suppose that the mathematical calculations give a certain outcome, or maybe several outcomes. These outcomes would then need to be interpreted in order to discover what they mean from a physical point of view. This ends the role of the mathematician, since this phase is maybe the most difficult engineering part of the process. It may happen that certain mathematical solutions have to be discarded since they contradict physical laws. Once the interpretation has been completed there is a return to the physical process, as the practical applicability of the results needs to be checked. In order to check these the quantities or functions that have been calculated are measured. The measurement is compared to the calculated result and in this way the physical model is validated. This validation may result in an adjustment of the physical model and another cycle in the loop is made. In this way the model is refined iteratively until we are satisfied about the validation. If there is a shortage of insight into the physical system, so that the physical model is not quite clear, measurements of the physical system may improve the physical model.

In the courses that are taught to students, models that have mainly been validated in this way are presented. However, it is important that students are aware of this process and the fact that the models that are presented may be a result of a difficult struggle for many years by several physicists, engineers and mathematicians. Sometimes students are given the opportunity to be involved in this process during research assignments.

1.3 THE CONCEPT OF A STOCHASTIC PROCESS

In probability theory a random variable is a rule that assigns a number to every outcome of an experiment, such as, for example, rolling a die. This random variable X is associated with a sample space S, such that according to a well-defined procedure to each event s in the

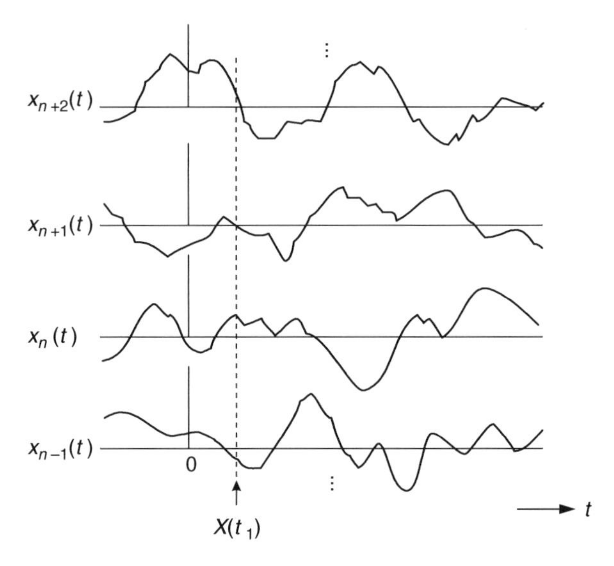

Figure 1.2 A few sample functions of a stochastic process

sample space a number is assigned to X and is denoted by $X(s)$. For stochastic processes, on the other hand, a time function $x(t, s)$ is assigned to every outcome in the sample space. Within the framework of the experiment the family (or ensemble) of all possible functions that can be realized is called the stochastic process and is denoted by $X(t, s)$. A specific waveform out of this family is denoted by $x_n(t)$ and is called a sample function or a realization of the stochastic process. When a realization in general is indicated the subscript n is omitted. Figure 1.2 shows a few sample functions that are supposed to constitute an ensemble. The figure gives an example of a finite number of possible realizations, but the ensemble may consist of an infinite number of realizations. The realizations may even be uncountable. A realization itself is sometimes called a stochastic process as well. Moreover, a stochastic process produces a random variable that arises from giving t a fixed value with s being variable. In this sense the random variable $X(t_1, s) = X(t_1)$ is found by considering the family of realizations at the fixed point in time t_1 (see Figure 1.2). Instead of $X(t_1)$ we will also use the notation X_1. The random variable X_1 describes the statistical properties of the process at the instant of time t_1. The expectation of X_1 is called the ensemble mean or the expected value or the mean of the stochastic process (at the instant of time t_1). Since t_1 may be arbitrarily chosen, the mean of the process will in general not be constant, i.e. it may have different values for different values of t. Finally, a stochastic process may represent a single number by giving both t and s fixed values. The phrase 'stochastic process' may therefore have four different interpretations. They are:

1. A family (or ensemble) of time functions. Both t and s are variables.
2. A single time function called a sample function or a realization of the stochastic process. Then t is a variable and s is fixed.
3. A random variable; t is fixed and s is variable.
4. A single number; both t and s are fixed.

Which of these four interpretations holds in a specific case should follow from the context.

Different classes of stochastic processes may be distinguished. They are classified on the basis of the characteristics of the realization values of the process x and the time parameter t. Both can be either continuous or discrete, in any combination. Based on this we have the following classes:

- Both the values of $X(t)$ and the time parameter t are continuous. Such a process is called a continuous stochastic process.
- The values of $X(t)$ are continuous, whereas time t is discrete. These processes are called discrete-time processes or continuous random sequences. In the remainder of the book we will use the term discrete-time process.
- If the values of $X(t)$ are discrete but the time axis is continuous, we call the process a discrete stochastic process.
- Finally, if both the process values and the time scale are discrete, we say that the process is a discrete random sequence.

In Table 1.1 an overview of the different classes of processes is presented. In order to get some feeling for stochastic processes we will consider a few examples.

Table 1.1 Summary of names of different processes

$X(t)$	Time	
	Continuous	Discrete
Continuous	Continuous stochastic process	Discrete-time process
Discrete	Discrete stochastic process	Discrete random sequence

1.3.1 Continuous Stochastic Processes

For this class of processes it is assumed that in principle the following holds:

$$-\infty < x(t) < \infty \quad \text{and} \quad -\infty < t < \infty \tag{1.1}$$

An example of this class was already given by Figure 1.2. This could be an ensemble of realizations of a thermal noise process as is, for instance, produced by a resistor, the characteristics of which are to be dealt with in Chapter 6. The underlying experiment is selecting a specific resistor from a collection of, let us say, 100 Ω resistors. The voltage across every selected resistor corresponds to one of the realizations in the figure.

Another example is given below.

Example 1.1:

The process we consider now is described by the equation

$$X(t) = \cos(\omega_0 t - \Theta) \tag{1.2}$$

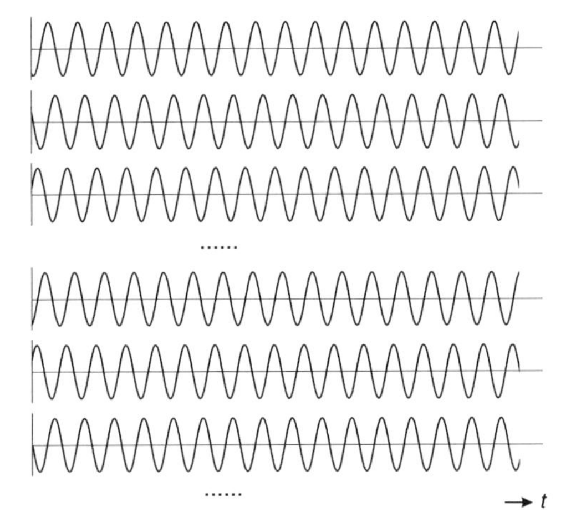

Figure 1.3 Ensemble of sample functions of the stochastic process $\cos(\omega_0 t - \Theta)$, with Θ uniformly distributed on the interval $(0, 2\pi]$

with ω_0 a constant and Θ a random variable with a uniform probability density function on the interval $(0, 2\pi]$. In this example the set of realizations is in fact uncountable, as Θ assumes continuous values. The ensemble of sample functions is depicted in Figure 1.3.

Thus each sample function consists of a cosine function with unity amplitude, but the phase of each sample function differs randomly from others. For each sample function a drawing is taken from the uniform phase distribution. We can imagine this process as follows. Consider a production process of crystal oscillators, all producing the same amplitude unity and the same radial frequency ω_0. When all those oscillators are switched on, their phases will be mutually independent. The family of all measured output waveforms can be considered as the ensemble that has been presented in Figure 1.3.

This process will get further attention in different chapters that follow.

□

1.3.2 Discrete-Time Processes (Continuous Random Sequences)

The description of this class of processes becomes more and more important due to the increasing use of modern digital signal processors which offer flexibility and increasing speed and computing power. As an example of a discrete-time process we can imagine sampling the process that was given in Figure 1.2. Let us suppose that to this process ideal sampling is applied at equidistant points in time with sampling period T_s; with ideal sampling we mean the sampling method where the values at T_s are replaced by delta functions of amplitude $X(nT_s)$ [6]. However, to indicate that it is now a discrete-time process we denote it by $X[n]$, where n is an integer running in principle from $-\infty$ to $+\infty$. We know from the sampling theorem (see Section 3.5.1 or, for instance, references [1] and [7]) that the original signal can perfectly be recovered from its samples, provided that the signals are band-limited. The process that is produced in this way is given in Figure 1.4, where the sample values are presented by means of the length of the arrows.

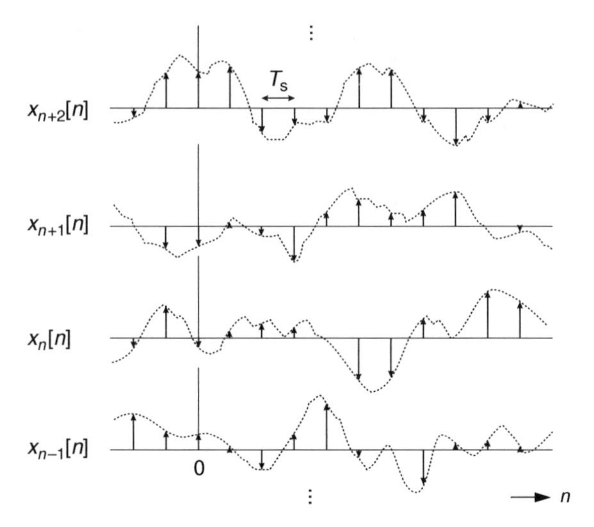

Figure 1.4 Example of a discrete-time stochastic process

Another important example of the discrete-time process is the so-called Poisson process, where there are no equidistant samples in time but the process produces 'samples' at random points in time. This process is an adequate model for shot noise and it is dealt with in Chapter 8.

1.3.3 Discrete Stochastic Processes

In this case the time is continuous and the values discrete. We present two examples of this class. The second one, the random data signal, is of great practical importance and we will consider it in further detail in Chapter 4.

Example 1.2:

This example is a very simple one. The ensemble of realizations consists of a set of constant time functions. According to the outcome of an experiment one of these constants may be chosen. This experiment can be, for example, the rolling of a die. In that case the number of realizations can be six $(n = 6)$, equal to the usual number of faces of a die. Each of the outcomes $s \in \{1, 2, 3, 4, 5, 6\}$ has a one-to-one correspondence to one of these numbered constant functions of time. The ensemble is depicted in Figure 1.5.

□

Example 1.3:

Another important stochastic process is the random data signal. It is a signal that is produced by many data sources and is described by

$$X(t) = \sum_n A_n p(t - nT - \Theta) \tag{1.3}$$

Figure 1.5 Ensemble of sample functions of the stochastic process constituted by a number of constant time functions

where $\{A_n\}$ are the data bits that are randomly chosen from the set $A_n \in \{+1, -1\}$. The rectangular pulse $p(t)$ of width T serves as the carrier of the information. Now Θ is supposed to be uniformly distributed on the bit interval $(0, T]$, so that all data sources of the family have the same bit period, but these periods are not synchronized. The ensemble is given in Figure 1.6.

□

1.3.4 Discrete Random Sequences

The discrete random sequence can be imagined to result from sampling a discrete stochastic process. Figure 1.7 shows the result of sampling the random data signal from Example 1.3.

We will base the further development of the concept, description and properties of stochastic processes on the continuous stochastic process. Then we will show how these are extended to discrete-time processes. The two other classes do not get special attention, but

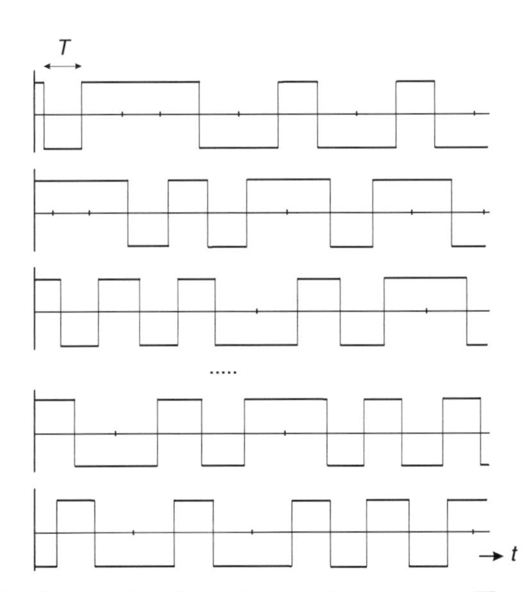

Figure 1.6 Ensemble of sample functions of the stochastic process $\sum_n A_n p(t - nT - \Theta)$, with Θ uniformly distributed on the interval $(0, T]$

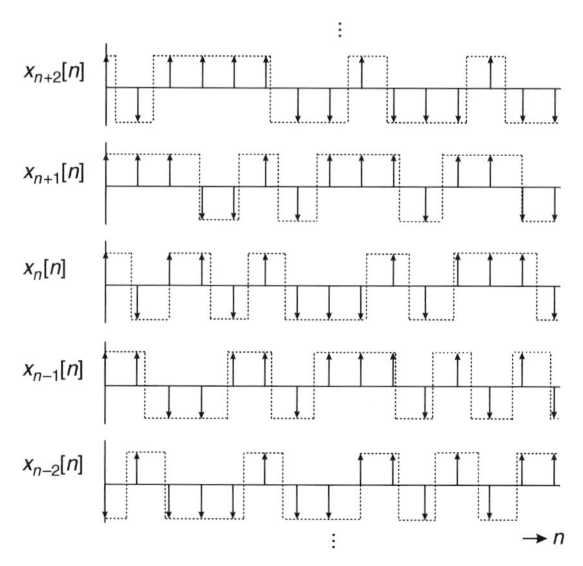

Figure 1.7 Example of a discrete random sequence

are considered as special cases of the former ones by limiting the realization values x to a discrete set.

1.3.5 Deterministic Function versus Stochastic Process

The concept of the stochastic process does not conflict with the theory of deterministic functions. It should be recognized that a deterministic function can be considered as nothing else but a special case of a stochastic process. This is elucidated by considering Example 1.1. If the random variable Θ is given the probability density function $f_\Theta(\theta) = \delta(\theta)$, then the stochastic process reduces to the function $\cos(\omega_0 t)$. The given probability density function is actually a discrete one with a single outcome. In fact, the ensemble of the process reduces in this case to a family comprising merely one member. This is a general rule; when the probability density function of the stochastic process that is governed by a single random variable consists of a single delta function, then a deterministic function results. This way of generalization avoids the often confusing discussion on the difference between a deterministic function on the one hand and a stochastic process on the other hand. In view of the consideration presented here they can actually be considered as members of the same class, namely the class of stochastic processes.

1.4 SUMMARY

Definitions of random signals and noise have been given. A random signal is, as a rule, an information carrying wanted signal that behaves randomly. Noise also behaves randomly but is unwanted and disturbs the signal. A common tool to describe both is the concept of a stochastic process. This concept has been explained and different classes of stochastic processes have been identified. They are distinguished by the behaviour of the time parameter and the values of the process. Both can either be continuous or discrete.

2

Stochastic Processes

In this chapter some basic concepts known from probability theory will be extended to include the time parameter. It is the time parameter that makes the difference between a random variable and a stochastic process. The basic concepts are: probability density function and correlation. The time dependence of the signals asks for a few new concepts, such as the correlation function, stationarity and ergodicity.

2.1 STATIONARY PROCESSES

As has been indicated in the introduction chapter we can fix the time parameter of a stochastic process. In this way we have a random variable, which can be characterized by means of a few deterministic numbers such as the mean, variance, etc. These quantities are defined using the probability density function. When fixing two time parameters we can consider two random variables simultaneously. Here also we can define joint random variables and, related to that, characterize quantities using the joint probability density function. In this way we can proceed, in general, to the case of N variables that are described by an N-dimensional joint probability density function, with N an arbitrary number.

Roughly speaking we can say that a stochastic process is stationary if its statistical properties do not depend on the time parameter. This rough definition will be elaborated in more detail in the rest of this chapter. There are several types of stationarity and for the main types we will present exact definitions in the sequel.

2.1.1 Cumulative Distribution Function and Probability Density Function

In order to be able to define stationarity, the probability distribution and density functions as they are applied to the stochastic process $X(t)$ have to be defined. For a fixed value of time parameter t_1 the cumulative probability distribution function or, for short, distribution function is defined by

$$F_X(x_1; t_1) \triangleq \mathrm{P}\{X(t_1) \leq x_1\} \tag{2.1}$$

Introduction to Random Signals and Noise W. van Etten

From this notation it follows that F_X may be a function of the value of t_1 that has been chosen.

For two random variables $X_1 = X(t_1)$ and $X_2 = X(t_2)$ we introduce the two-dimensional extension of Equation (2.1):

$$F_X(x_1, x_2; t_1, t_2) \triangleq \mathrm{P}\{X(t_1) \leq x_1, X(t_2) \leq x_2\} \tag{2.2}$$

the second-order, joint probability distribution function. In an analogous way we denote the Nth-order, joint probability distribution function

$$F_X(x_1, \ldots, x_N; t_1, \ldots, t_N) \triangleq \mathrm{P}\{X(t_1) \leq x_1, \ldots, X(t_N) \leq x_N\} \tag{2.3}$$

The corresponding (joint) probability density functions are found by taking the derivatives respectively of Equations (2.1) to (2.3):

$$f_X(x_1; t_1) \triangleq \frac{\partial F_X(x_1; t_1)}{\partial x_1} \tag{2.4}$$

$$f_X(x_1, x_2; t_1, t_2) \triangleq \frac{\partial^2 F_X(x_1, x_2; t_1, t_2)}{\partial x_1 \partial x_2} \tag{2.5}$$

$$f_X(x_1, \ldots, x_N; t_1, \ldots, t_N) \triangleq \frac{\partial^N F_X(x_1, \ldots, x_N; t_1, \ldots, t_N)}{\partial x_1 \cdots \partial x_N} \tag{2.6}$$

Two processes $X(t)$ and $Y(t)$ are called statistically independent if the set of random variables $\{X(t_1), X(t_2), \ldots, X(t_N)\}$ is independent of the set of random variables $\{Y(t'_1), Y(t'_2), \ldots, Y(t'_M)\}$, for each arbitrary choice of the time parameters $\{t_1, t_2, \ldots, t_N; t'_1, t'_2, \ldots, t'_M\}$. Independence implies that the joint probability density function can be factored in the following way:

$$\begin{aligned} &f_{X,Y}(x_1, \ldots, x_N; y_1, \ldots, y_M; t_1, \ldots, t_N; t'_1, \ldots, t'_M) \\ &\quad = f_X(x_1, \ldots, x_N; t_1, \ldots, t_N) \cdot f_Y(y_1, \ldots, y_M; t'_1, \ldots, t'_M) \end{aligned} \tag{2.7}$$

Thus, the joint probability density function of two independent processes is written as the product of the two marginal probability density functions.

2.1.2 First-Order Stationary Processes

A stochastic process is called a first-order stationary process if the first-order probability density function is independent of time. Mathematically this can be stated as

$$f_X(x_1; t_1) = f_X(x_1; t_1 + \tau) \tag{2.8}$$

holds for all τ. As a consequence of this property the mean value of such a process, denoted by $\overline{X(t)}$, is

$$\overline{X(t)} \equiv \mathrm{E}[X(t)] \triangleq \int x\, f_X(x; t)\, \mathrm{d}x = \text{constant} \tag{2.9}$$

i.e. it is independent of time.

2.1.3 Second-Order Stationary Processes

A stochastic process is called a second-order stationary process if for the two-dimensional joint probability density function

$$f_X(x_1, x_2; t_1, t_2) = f_X(x_1, x_2; t_1 + \tau, t_2 + \tau) \tag{2.10}$$

for all τ. It is easy to verify that Equation (2.10) is only a function of the time difference $t_2 - t_1$ and does not depend on the absolute time. In order to gain that insight put $\tau = -t_1$.

A process that is second-order stationary is first-order stationary as well, since the second-order joint probability density function uniquely determines the lower-order (in this case first-order) probability density function.

2.1.4 *N*th-Order Stationary Processes

By extending the reasoning from the last subsection to N random variables $X_i = X(t_i)$, for $i = 1, \ldots, N$, we arrive at an Nth-order stationary process. The Nth-order joint probability density function is once more independent of a time shift; i.e.

$$f_X(x_1, \ldots, x_N; t_1, \ldots, t_N) = f_X(x_1, \ldots, x_N; t_1 + \tau, \ldots, t_N + \tau) \tag{2.11}$$

for all τ. A process that is Nth-order stationary is stationary to all orders $k \leq N$. An Nth-order stationary process where N can have an arbitrary large value is called a strict-sense stationary process.

2.2 CORRELATION FUNCTIONS

2.2.1 The Autocorrelation Function, Wide-Sense Stationary Processes and Ergodic Processes

The autocorrelation function of a stochastic process is defined as the correlation $\mathrm{E}[X_1 X_2]$ of the two random variables $X_1 = X(t_1)$ and $X_2 = X(t_2)$. These random variables are achieved by considering all realization values of the stochastic process at the instants of time t_1 and t_2 (see Figure 2.1). In general it will be a function of these two times instants. The autocorrelation function is denoted as

$$R_{XX}(t_1, t_2) \triangleq \mathrm{E}[X(t_1)X(t_2)] = \iint x_1 x_2 \, f_X(x_1, x_2; t_1, t_2) \, \mathrm{d}x_1 \, \mathrm{d}x_2 \tag{2.12}$$

Substituting $t_1 = t$ and $t_2 = t_1 + \tau$, Equation (2.12) becomes

$$R_{XX}(t, t + \tau) = \mathrm{E}[X(t)\,X(t + \tau)] \tag{2.13}$$

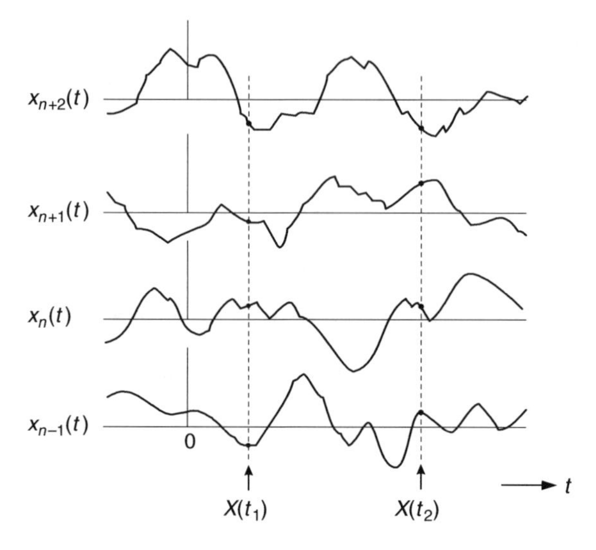

Figure 2.1 The autocorrelation of a stochastic process by considering $E[X(t_1)X(t_2)]$

Since for a second-order stationary process the two-dimensional joint probability density function depends only on the time difference, the autocorrelation function will also be a function of the time difference τ. Then Equation (2.13) can be written as

$$R_{XX}(t, t+\tau) = R_{XX}(\tau) \tag{2.14}$$

The mean and autocorrelation function of a stochastic process are often its most characterizing features. Mostly, matters become easier if these two quantities do not depend on absolute time. A second-order stationary process guarantees this independence but at the same time places severe demands on the process. Therefore we define a broader class of stochastic processes, the so-called wide-sense stationary processes.

Definition

A process $X(t)$ is called wide-sense stationary if it satisfies the conditions

$$\begin{aligned} E[X(t)] &= \overline{X(t)} = \text{constant} \\ E[X(t)\,X(t+\tau)] &= R_{XX}(\tau) \end{aligned} \tag{2.15}$$

It will be clear that a second-order stationary process is also wide-sense stationary. The converse, however, is not necessarily true.

Properties of $R_{XX}(\tau)$

If a process is at least wide-sense stationary then its autocorrelation function exhibits the following properties:

1. $|R_{XX}(\tau)| \leq R_{XX}(0)$ (2.16)
   i.e. $|R_{XX}(\tau)|$ attains its maximum value for $\tau = 0$.
2. $R_{XX}(-\tau) = R_{XX}(\tau)$ (2.17)
   i.e. $R_{XX}(\tau)$ is an even function of τ.
3. $R_{XX}(0) = \mathrm{E}[X^2(t)]$ (2.18)
4. If $X(t)$ has no periodic component then $R_{XX}(\tau)$ comprises a constant term equal to $\overline{X(t)}^2$, i.e. $\lim_{|\tau|\to\infty} R_{XX}(\tau) = \overline{X(t)}^2$.
5. If $X(t)$ has a periodic component then $R_{XX}(\tau)$ will comprise a periodic component as well, and which has the same periodicity.

A function that does not satisfy these properties cannot be the autocorrelation function of a wide-sense stationary process. It will be clear from properties 1 and 2 that $R_{XX}(\tau)$ is not allowed to exhibit an arbitrary shape.

Proofs of the properties:

1. To prove property 1 let us consider the expression

$$\begin{aligned} \mathrm{E}[\{X(t) \pm X(t+\tau)\}^2] &= \mathrm{E}[X^2(t) + X^2(t+\tau) \pm 2X(t)\,X(t+\tau)] \\ &= 2\{R_{XX}(0) \pm R_{XX}(\tau)\} \geq 0 \end{aligned} \tag{2.19}$$

   Since the expectation $\mathrm{E}[\{X(t) \pm X(t+\tau)\}^2]$ is taken over the squared value of a certain random variable, this expectation should be greater than or equal to zero. From the last line of Equation (2.19) property 1 is concluded.

2. The proof of property 2 is quite simple. In the definition of the autocorrelation function substitute $t' = t + \tau$ and the proof proceeds as follows:

$$\begin{aligned} R_{XX}(\tau) &= \mathrm{E}[X(t)\,X(t+\tau)] = \mathrm{E}[X(t'-\tau)\,X(t')] \\ &= \mathrm{E}[X(t')\,X(t'-\tau)] = R_{XX}(-\tau) \end{aligned} \tag{2.20}$$

3. Property 3 follows immediately from the definition of $R_{XX}(\tau)$ by inserting $\tau = 0$.
4. From a physical point of view most processes have the property that the random variables $X(t)$ and $X(t+\tau)$ are independent when $\tau \to \infty$. Invoking once more the definition of the

autocorrelation function it follows that

$$\lim_{\tau \to \infty} R_{XX}(\tau) = \lim_{\tau \to \infty} \mathrm{E}[X(t)\, X(t+\tau)]$$
$$= \mathrm{E}[X(t)]\, \mathrm{E}[X(t+\tau)] = \mathrm{E}^2[X(t)] = \overline{X}^2 \tag{2.21}$$

5. Periodic processes may be decomposed into cosine and sine components according to Fourier analysis. It therefore suffices to consider the autocorrelation function of one such component:

$$\mathrm{E}[\cos(\omega t - \Theta)\, \cos(\omega t + \omega\tau - \Theta)] = \tfrac{1}{2}\mathrm{E}[\cos(\omega\tau) + \cos(2\omega t + \omega\tau - 2\Theta)] \tag{2.22}$$

   Since our considerations are limited to wide-sense stationary processes, the autocorrelation function should be independent of the absolute time t, and thus the expectation of the last term of the latter expression should be zero. Thus only the term comprising $\cos(\omega\tau)$ remains after taking the expectation, which proves property 5.

When talking about the mean or expectation (denoted by $\mathrm{E}[\cdot]$) the statistical average over the ensemble of realizations is meant. Since stochastic processes are time functions we can define another average, namely the time average, given by

$$\mathrm{A}[X(t)] \triangleq \lim_{T \to \infty} \frac{1}{2T} \int_{-T}^{T} x(t)\mathrm{d}t \tag{2.23}$$

When taking this time average only one single sample function can be involved; consequently, expressions like $\mathrm{A}[X(t)]$ and $\mathrm{A}[X(t)\, X(t+\tau)]$ will be random variables.

Definition

A wide-sense stationary process $X(t)$ satisfying the two conditions

$$\mathrm{A}[X(t)] = \mathrm{E}[X(t)] = \overline{X(t)} \tag{2.24}$$
$$\mathrm{A}[X(t)\, X(t+\tau)] = \mathrm{E}[X(t)\, X(t+\tau)] = R_{XX}(\tau) \tag{2.25}$$

is called an ergodic process.

In other words, an ergodic process has time averages $\mathrm{A}[X(t)]$ and $\mathrm{A}[X(t)\, X(t+\tau)]$ that are non-random because these time averages equal the ensemble averages $\overline{X(t)}$ and $R_{XX}(\tau)$. In the same way as several types of stationary process can be defined, several types of ergodic processes may also be introduced [1]. We will confine ourselves to the forms defined by the Equations (2.24) and (2.25). Ergodicity puts more severe demands on the process than stationarity and it is often hard to prove that indeed a process is ergodic; often it is impossible. In practice ergodicity is often just assumed without proof, unless the opposite is evident. In most cases there is no alternative, as one does not have access to the entire family (ensemble) of sample functions, but rather just to one or a few members of it, for example one resistor, transistor or comparable noisy device is available. By assuming ergodicity a number of important statistical properties, such as the mean and the autocorrelation function of a process may be estimated from the observation of a single available realization. Fortunately, it appears that many processes are ergodic, but one should always be aware that at times one can encounter a process that is not ergodic. Later in this chapter we will develop a test for a certain class of ergodic processes.

Example 2.1:

As an example consider the process $X(t) = A\cos(\omega t - \Theta)$, with A a constant amplitude, ω a fixed but arbitrary radial frequency and Θ a random variable that is uniformly distributed on the interval $(0, 2\pi]$. The question is whether this process is ergodic in the sense as defined by Equations (2.24) and (2.25). To answer this we determine both the ensemble mean and the time average. For the time average it is found that

$$\mathrm{A}[X(t)] = \lim_{T\to\infty} \frac{1}{2T} A \int_{-T}^{T} \cos(\omega t - \Theta)\,\mathrm{d}t = \lim_{T\to\infty} \frac{1}{2T} A \frac{1}{\omega} \sin(\omega t - \Theta)\Big|_{-T}^{T} = 0 \tag{2.26}$$

The ensemble mean is

$$\mathrm{E}[X(t)] = \int f_\Theta(\theta)\, A\cos(\omega t - \theta)\,\mathrm{d}\theta = \frac{1}{2\pi} A \int_0^{2\pi} \cos(\omega t - \theta)\,\mathrm{d}\theta = 0 \tag{2.27}$$

Hence, time and ensemble averages are equal.

Let us now calculate the two autocorrelation functions. For the time-averaged autocorrelation function it is found that

$$\begin{aligned}\mathrm{A}[X(t)\,X(t+\tau)] &= \lim_{T\to\infty} \frac{1}{2T} A^2 \int_{-T}^{T} \cos(\omega t - \theta)\,\cos(\omega t + \omega\tau - \theta)\,\mathrm{d}t \\ &= \lim_{T\to\infty} \frac{1}{2T} \frac{1}{2} A^2 \int_{-T}^{T} [\cos(2\omega t + \omega\tau - 2\theta) + \cos(\omega\tau)]\,\mathrm{d}t\end{aligned} \tag{2.28}$$

The first term of the latter integral equals 0. The second term of the integrand does not depend on the integration variable. Hence, the autocorrelation function is given by

$$\mathrm{A}[X(t)\,X(t+\tau)] = \tfrac{1}{2} A^2 \cos\omega\tau \tag{2.29}$$

Next we consider the statistical autocorrelation function

$$\begin{aligned}\mathrm{E}[X(t)\,X(t+\tau)] &= \frac{1}{2\pi} A^2 \int_0^{2\pi} \cos(\omega t - \theta)\,\cos(\omega t + \omega\tau - \theta)\,\mathrm{d}\theta \\ &= \frac{1}{2\pi}\frac{1}{2} A^2 \int_0^{2\pi} [\cos(2\omega t + \omega\tau - 2\theta) + \cos(\omega\tau)]\,\mathrm{d}\theta\end{aligned} \tag{2.30}$$

Of the latter integral the first part is 0. Again, the second term of the integrand does not depend on the integration variable. The autocorrelation function is therefore

$$\mathrm{E}[X(t)\,X(t+\tau)] = \tfrac{1}{2}A^2 \cos\omega\tau \tag{2.31}$$

Hence both first-order means (time average and statistical mean) and second-order means (time-averaged and statistical autocorrelation functions) are equal. It follows that the process is ergodic.

The process $\cos(\omega t - \Theta)$ with $f_\Theta(\theta) = \delta(\theta)$ equals the deterministic function $\cos\omega t$. This process is not ergodic, since it is easily verified that the expectation (in this case the function itself) is time-dependent and thus not stationary, which is a condition for ergodicity. This example, where a probability density function that consists of a δ function reduces the process to a deterministic function, has also been mentioned in Chapter 1.

□

2.2.2 Cyclo-Stationary Processes

A process $X(t)$ is called cyclo-stationary (or periodically stationary) if the probability density function is independent of a shift in time over an integer multiple of a constant value T (the period time), so that

$$f_X(x_1,\ldots,x_N;t_1,\ldots,t_N) = f_X(x_1,\ldots,x_N;t_1+mT,\ldots,t_N+mT) \tag{2.32}$$

for each integer value of m. A cyclo-stationary process is not stationary, since Equation (2.11) is not valid for all values of τ, but only for discrete values $\tau = mT$. However, the discrete-time process $X(mT+\tau)$ is stationary for all values of τ.

A relation exists between cyclo-stationary processes and stationary processes. To see this relation it is evident from Equation (2.32) that

$$F_X(x_1,\ldots,x_N;t_1,\ldots,t_N) = F_X(x_1,\ldots,x_N;t_1+mT,\ldots,t_N+mT) \tag{2.33}$$

Next consider the modified process $\underline{X}(t) = X(t-\Theta)$, where $X(t)$ is cyclo-stationary and Θ a random variable that has a uniform probability density function on the period interval $(0,T]$. Now we define the event $\mathscr{A}$ as

$$\mathscr{A} = \{\underline{X}(t_1+\tau) \le x_1,\ldots,\underline{X}(t_N+\tau) \le x_N\} \tag{2.34}$$

The probability that this event will occur is

$$\mathrm{P}(\mathscr{A}) = \int_0^T \mathrm{P}(\mathscr{A}|\Theta=\theta)\,f_\Theta(\theta)\,\mathrm{d}\theta = \frac{1}{T}\int_0^T \mathrm{P}(\mathscr{A}|\Theta=\theta)\,\mathrm{d}\theta \tag{2.35}$$

For the latter integrand we write

$$\begin{aligned}\mathrm{P}(\mathscr{A}|\Theta=\theta) &= \mathrm{P}\{X(t_1+\tau-\theta) \le x_1,\ldots,X(t_N+\tau-\theta) \le x_N\}\\ &= F_X(x_1,\ldots,x_N;t_1+\tau-\theta,\ldots,t_N+\tau-\theta)\end{aligned} \tag{2.36}$$

Substituting this result in Equation (2.35) yields

$$\mathrm{P}(\mathscr{A}) = \frac{1}{T}\int_0^T F_X(x_1, \ldots, x_N; t_1 + \tau - \theta, \ldots, t_N + \tau - \theta)\,\mathrm{d}\theta \tag{2.37}$$

As $X(t)$ is cyclo-stationary Equation (2.37) is independent of τ. From Equation (2.34) it follows, therefore, that $\mathrm{P}(\mathscr{A})$ represents the probability distribution function of the process $\underline{X}(t)$. Thus we have the following theorem.

Theorem 1

If $X(t)$ is a cyclo-stationary process with period time T and Θ is a random variable that is uniformly distributed on the interval $(0, T]$, then the process $\underline{X}(t) = X(t - \Theta)$ is stationary with the probability distribution function

$$F_{\underline{X}}(x_1, \ldots, x_N; t_1, \ldots, t_N) = \frac{1}{T}\int_0^T F_X(x_1, \ldots, x_N; t_1 - \tau, \ldots, t_N - \tau)\,\mathrm{d}\tau \tag{2.38}$$

A special case consists of the situation where $X(t) = p(t)$ is a deterministic, periodic function with period T. Then, as far as the first-order probability distribution function $F_{\underline{X}}(x)$ is concerned, the integral from Equation (2.38) can be interpreted as the relative fraction of time during which $X(t)$ is smaller or equal to x. This is easily understood when realizing that for a deterministic function $F_X(x_1; t_1)$ is either zero or one, depending on whether $p(t_1)$ is larger or smaller than x_1.

If we take $X(t) = p(t)$, then this process $X(t)$ is strict sense cyclo-stationary and from the foregoing we have the following theorem.

Theorem 2

If $\underline{X}(t) = p(t - \Theta)$ is an arbitrary, periodic waveform with period T and Θ a random variable that is uniformly distributed on the interval $(0, T]$, then the process $\underline{X}(t)$ is strict-sense stationary and ergodic. The probability distribution function of this process reads

$$F_{\underline{X}}(x_1, \ldots, x_N; t_1, \ldots, t_N) = \frac{1}{T}\int_0^T F_p(p_1, \ldots, p_N; t_1 - \tau, \ldots, t_N - \tau)\,\mathrm{d}\tau \tag{2.39}$$

The mean value of the process equals

$$\mathrm{E}[\underline{X}(t)] = \frac{1}{T}\int_0^T p(t)\,\mathrm{d}t = \mathrm{A}[p(t)] \tag{2.40}$$

and the autocorrelation function

$$R_{\underline{XX}}(\tau) = \frac{1}{T}\int_0^T p(t)\,p(t + \tau)\,\mathrm{d}t = \mathrm{A}[p(t)\,p(t + \tau)] \tag{2.41}$$

This latter theorem is a powerful expedient when proving strict-sense stationarity and ergodicity of processes that often occur in practice. In such cases the probability distribution function is found by means of the integral given by Equation (2.39). For this integral the same interpretation is valid as for that from Equation (2.38). From the probability distribution function the probability density function can be derived using Equation (2.4). By adding a random phase Θ , with Θ uniformly distributed on the interval $(0, T]$, to a cyclo-stationary process the process can be made stationary. Although this seems to be an artificial operation, it is not so from a physical point of view. If we imagine that the ensemble of realizations originates from a set of signal generators, let us say sinusoidal wave generators, all of them tuned to the same frequency, then the waves produced by these generators will as a rule not be synchronized in phase.

Example 2.2:

The process $X(t) = \cos(\omega t)$ is not stationary, its mean value being $\cos(\omega t)$; however, it is cyclo-stationary. On the contrary, the modified process $\underline{X}(t) = \cos(\omega t - \Theta)$, with Θ uniformly distributed on the interval $(0, 2\pi]$, is strict-sense stationary and ergodic, based on Theorem 2. The ergodicity of this process was already concluded when dealing with Example 2.1. Moreover, we derived the autocorrelation function of this process as $\frac{1}{2}\cos(\omega\tau)$.

Let us now elaborate this example. We will derive the probability distribution and density functions, based on Theorem 2. For this purpose remember that the probability distribution function is given by Equation (2.39) and that this integral is interpreted as the relative fraction of time during which $X(t)$ is smaller than or equal to x. This interpretation has been further explained by means of Figure 2.2. In this figure one complete period of a cosine is presented. The constant value x is indicated. The duration that the given realization is smaller than or equal to x has been drawn by means of the bold line pieces, which are indicated by T_1 and T_2, and the complete second half of the cycle, which has a length of π. It can be seen that the line pieces T_1 and T_2 are of equal length, namely $\arcsin x$. Finally, the probability distribution function is found by adding all the bold line pieces and dividing the result by the period 2π. This leads to the probability distribution function

$$F_{\underline{X}}(x) = \mathrm{P}\{\underline{X}(t) \leq x\} = \begin{cases} \frac{1}{2} + \frac{1}{\pi}\arcsin x, & |x| \leq 1 \\ 0, & x < -1 \\ 1, & x > 1 \end{cases} \tag{2.42}$$

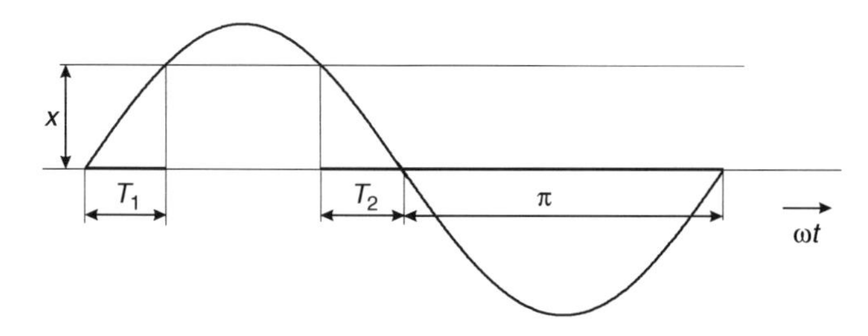

Figure 2.2 Figure to help determine the probability distribution function of the random phased cosine

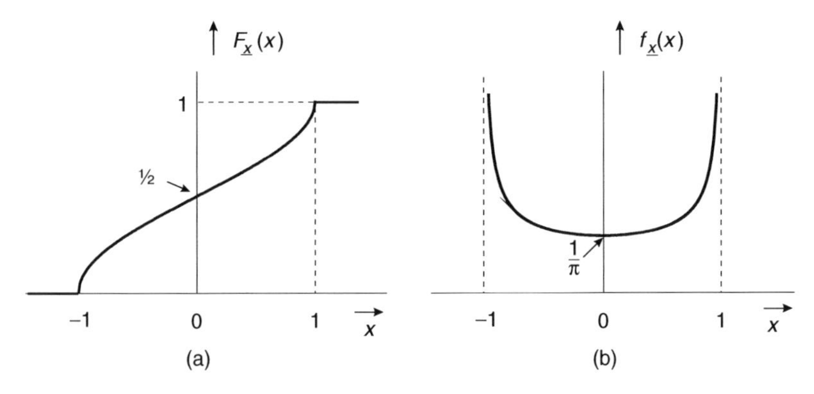

Figure 2.3 (a) The probability distribution function and (b) the probability density function of the random phased cosine

This function is depicted in Figure 2.3(a). From the probability distribution function the probability density function is easily derived by taking the derivative, i.e.

$$f_{\underline{X}}(x) \triangleq \frac{\mathrm{d}F_{\underline{X}}(x)}{\mathrm{d}x} = \begin{cases} \frac{1}{\pi}\frac{1}{\sqrt{1-x^2}}, & |x| \leq 1 \\ 0, & |x| > 1 \end{cases} \tag{2.43}$$

This function has been plotted in Figure 2.3(b). Note the asymptotic values of the function for both $x = 1$ and $x = -1$.

□

Example 2.3:

The random data signal

$$X(t) = \sum_n A_n p(t - nT) \tag{2.44}$$

with A_n a stationary sequence of binary random variables that are selected out of the set $\{-1, +1\}$ and with autocorrelation sequence

$$\mathrm{E}[A_n A_k] = \mathrm{E}[A_n A_{n+m}] = \mathrm{E}[A_n A_{n-m}] = R_m \tag{2.45}$$

constitutes a cyclo-stationary process, where Theorems 1 and 2 can be applied. Properties of this random data signal will be derived in more detail later on (see Section 4.5).

□

2.2.3 The Cross-Correlation Function

The cross-correlation function of two stochastic processes $X(t)$ and $Y(t)$ is defined as

$$R_{XY}(t, t+\tau) \triangleq \mathrm{E}[X(t)\, Y(t+\tau)] \tag{2.46}$$

$X(t)$ and $Y(t)$ are jointly wide-sense stationary if both $X(t)$ and $Y(t)$ are wide-sense stationary and if the cross-correlation function $R_{XY}(t, t+\tau)$ is independent of the absolute time parameter, i.e.

$$R_{XY}(t, t+\tau) = \mathrm{E}[X(t)\,Y(t+\tau)] = R_{XY}(\tau) \tag{2.47}$$

Properties of $R_{XY}(\tau)$

If two processes $X(t)$ and $Y(t)$ are jointly wide-sense stationary, then the cross-correlation function has the following properties:

1. $R_{XY}(-\tau) = R_{YX}(\tau)$ (2.48)
2. $|R_{XY}(\tau)| \le \sqrt{R_{XX}(0)\,R_{YY}(0)}$ (2.49)
3. $|R_{XY}(\tau)| \le \frac{1}{2}[R_{XX}(0) + R_{YY}(0)]$ (2.50)

A function that does not satisfy these properties cannot be the cross-correlation function of two jointly wide-sense stationary processes.

Proofs of the properties:

1. Property 1 is proved as follows. In the definition of the cross-correlation function replace $t-\tau$ by t' and do some manipulation as shown below:

$$\begin{aligned} R_{XY}(-\tau) &= \mathrm{E}[X(t)\,Y(t-\tau)] = \mathrm{E}[X(t'+\tau)\,Y(t')] \\ &= \mathrm{E}[Y(t')\,X(t'+\tau)] = R_{YX}(\tau) \end{aligned} \tag{2.51}$$

2. To prove property 2 we consider the expectation of the process $\{X(t) + cY(t+\tau)\}^2$, where c is a constant; i.e. we investigate

$$\begin{aligned} \mathrm{E}[\{X(t) + cY(t+\tau)\}^2] &= \mathrm{E}[X^2(t) + c^2Y^2(t+\tau) + 2cX(t)\,Y(t+\tau)] \\ &= \mathrm{E}[X^2(t)] + c^2\mathrm{E}[Y^2(t+\tau)] + 2c\mathrm{E}[X(t)\,Y(t+\tau)] \\ &= c^2R_{YY}(0) + 2cR_{XY}(\tau) + R_{XX}(0) \end{aligned} \tag{2.52}$$

This latter expression is a quadratic form as a function of c, and since it is the expectation of a quantity squared, this expression can never be less than zero. As a consequence the discriminant cannot be positive; i.e.

$$R_{XY}^2(\tau) - R_{XX}(0)\,R_{YY}(0) \le 0 \tag{2.53}$$

From this property 2 follows.

3. Property 3 is a consequence of the well-known fact that the arithmetic mean of two positive numbers is always greater than or equal to their geometric mean.

If for two processes $X(t)$ and $Y(t)$

$$R_{XY}(t, t+\tau) = 0, \quad \text{for all } t \text{ and } \tau \tag{2.54}$$

then we say that $X(t)$ and $Y(t)$ are orthogonal processes. In case two processes $X(t)$ and $Y(t)$ are statistically independent, the cross-correlation function can be written as

$$R_{XY}(t, t+\tau) = \mathrm{E}[X(t)]\,\mathrm{E}[Y(t+\tau)] \tag{2.55}$$

If, moreover, $X(t)$ and $Y(t)$ are at least wide-sense stationary, then Equation (2.55) becomes

$$R_{XY}(t, t+\tau) = \overline{X}\,\overline{Y} \tag{2.56}$$

Two stochastic processes $X(t)$ and $Y(t)$ are called jointly ergodic if the individual processes are ergodic and if the time-averaged cross-correlation function equals the statistical cross-correlation function, i.e. if

$$\mathrm{A}[X(t)\,Y(t+\tau)] = \mathrm{E}[X(t)\,Y(t+\tau)] = R_{XY}(\tau) \tag{2.57}$$

In practice one more often uses spectra, to be dealt with in the next chapter, than correlation functions, as the measurement equipment for spectra is more developed than that for correlations. In that chapter it will be shown that the correlation function acts as the basis for calculating the spectrum. However, the correlation function in itself also has interesting applications, as is concluded from the following examples.

Example 2.4:

It will be shown that based on a correlation function, by means of the system described in this example, one is able to measure a distance. Consider a system (see Figure 2.4) where a signal source produces a random signal, being a realization of a stochastic process. Let us

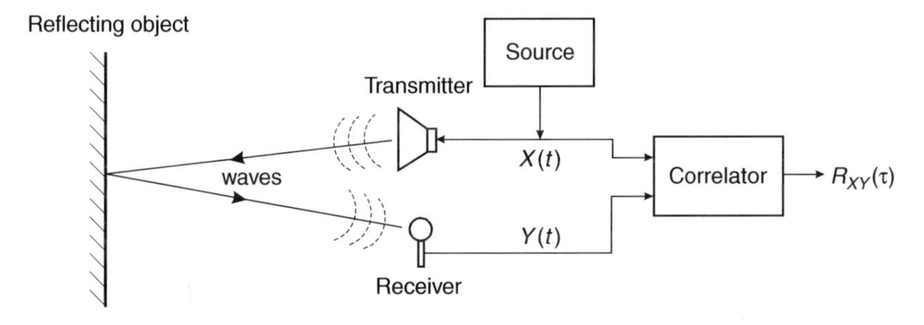

Figure 2.4 Set-up for measuring a distance based on the correlation function

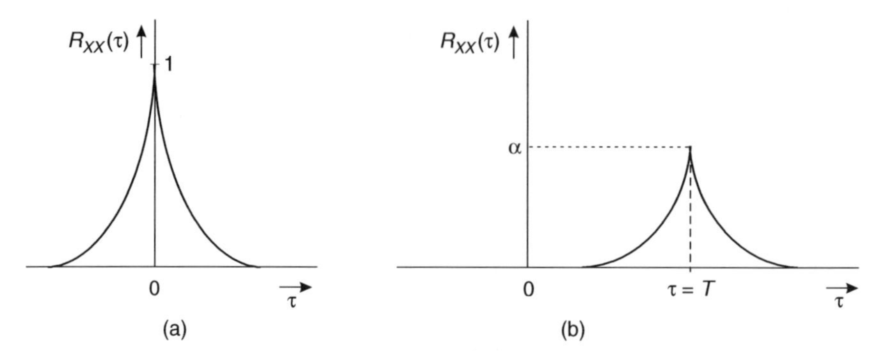

Figure 2.5 (a) The autocorrelation function of the transmitted signal and (b) the measured cross-correlation function of the distance measuring set-up

assume the process to be wide-sense stationary. The signal is applied to a transmitter that produces a wave in a transmission medium; let it be an acoustic wave or an electromagnetic wave. We denote the transmitted random wave by $X(t)$. Let us further suppose that the transmitted wave strikes a distant object and that this object (partly) reflects the wave. Then this reflected wave will travel backwards to the position of the measuring equipment. The measuring equipment comprises a receiver and the received signal is denoted as $Y(t)$. Both the transmitted signal $X(t)$ and the received signal $Y(t)$ are applied to a correlator that produces the cross-correlation function $R_{XY}(\tau)$. In the next section it will be explained how this correlation equipment operates. The reflected wave will be a delayed and attenuated version of the transmitted wave; i.e. we assume $Y(t) = \alpha X(t - T)$, where T is the total travel time. The cross-correlation function will be

$$R_{XY}(\tau) = \mathrm{E}[X(t)\, Y(t + \tau)] = \mathrm{E}[X(t)\, \alpha X(t - T + \tau)] = \alpha R_{XX}(\tau - T) \tag{2.58}$$

Most autocorrelation functions have a peak at $\tau = 0$, as shown in Figure 2.5(a). Let us normalize this peak to unity; then the cross-correlation result will be as depicted in Figure 2.5(b). From this latter picture a few conclusions may be drawn with respect to the application at hand. Firstly, when we detect the position of the peak in the cross-correlation function we will be able to establish T and if the speed of propagation of the wave in the medium is known, then the distance of the object can be derived from that. Secondly, the relative height α of the peak can be interpreted as a measure for the size of the object.

It will be clear that this method is very useful in such ranging systems as radar and underwater acoustic distance measurement. Most ranging systems use pulsed continuous wave (CW) signals for that. The advantage of the system presented here is the fact that for the transmitted signal a noise waveform is used. Such a waveform cannot easily be detected by the probed object, in contrast to the pulsed CW systems, since it has no replica available of the transmitted signal and therefore is not able to perform the correlation. The probed object only observes an increase in received noise level.

□

Example 2.5:

Yet another interesting example of the application of the correlation concept is in the field of reducing the distortion of received information signals. Let us suppose that a private subscriber has on the roof of his house an antenna for receiving TV broadcast signals. Due to a tall building near his house the TV signal is reflected, so that the subscriber receives the signal from a certain transmitter twice, once directly from the transmitter and a second time reflected from the neighbouring building. On the TV screen this produces a ghost of the original picture and spoils the picture. We call the direct signal $X(t)$ and the reflected one will then be $\alpha X(t-T)$, where T represents the difference in travel time between the direct and the reflected signal. The total received signal is therefore written as $Y(t) = X(t) + \alpha X(t-T)$. Let us consider the autocorrelation function of this process:

$$\begin{aligned} R_{YY}(\tau) &= \mathrm{E}[\{X(t) + \alpha X(t-T)\}\,\{X(t+\tau) + \alpha X(t-T+\tau)\}] \\ &= \mathrm{E}[X(t)\,X(t+\tau) + \alpha X(t)\,X(t-T+\tau) + \alpha X(t-T)\,X(t+\tau) \\ &\quad + \alpha^2 X(t-T)\,X(t-T+\tau)] \\ &= (1+\alpha^2)R_{XX}(\tau) + \alpha R_{XX}(\tau-T) + \alpha R_{XX}(\tau+T) \end{aligned} \tag{2.59}$$

The autocorrelation function $R_{YY}(\tau)$ of the received signal $Y(t)$ will consist of that of the original signal $R_{XX}(\tau)$ multiplied by $1+\alpha^2$ and besides that two shifted versions of $R_{XX}(\tau)$. These versions are multiplied by α and shifted in time over respectively T and $-T$. If it is assumed that the autocorrelation function $R_{XX}(\tau)$ shows the same peaked appearance as in the previous example, then the autocorrelation function of the received signal $Y(t)$ looks like that in Figure 2.6. Let us once again suppose that both α and T can be determined from this measurement. Then we will show that these parameters can be used to reduce the distortion caused by the reflection from the nearby building; namely we delay the received signal by an amount of T and multiply this delayed version by α. This delayed and multiplied version is subtracted from the received signal, so that after this operation we have the signal $Z(t) = Y(t) - \alpha Y(t-T)$. Inserting the undistorted signal $X(t)$ into this yields

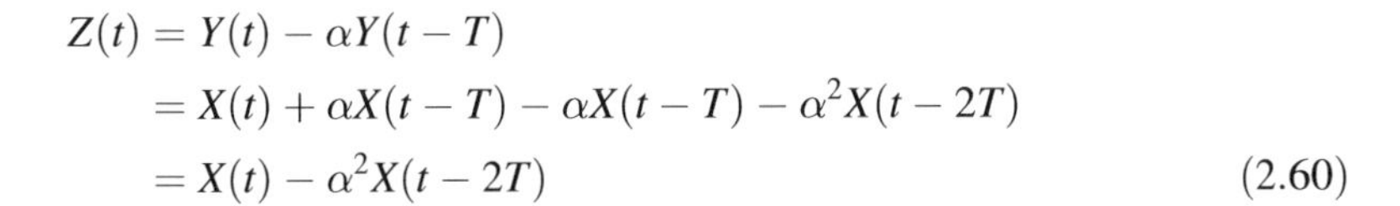

$$\begin{aligned} Z(t) &= Y(t) - \alpha Y(t-T) \\ &= X(t) + \alpha X(t-T) - \alpha X(t-T) - \alpha^2 X(t-2T) \\ &= X(t) - \alpha^2 X(t-2T) \end{aligned} \tag{2.60}$$

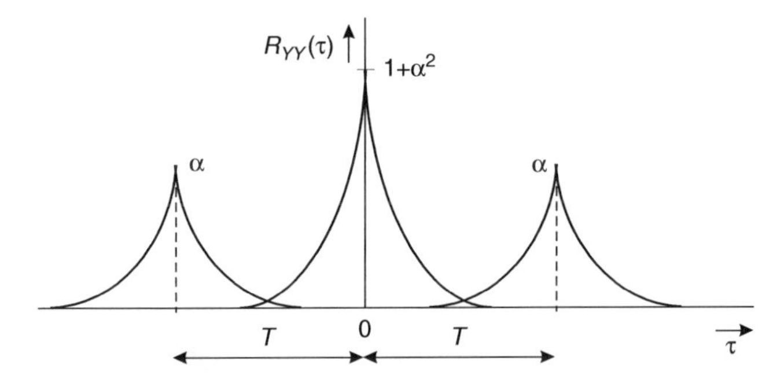

Figure 2.6 The autocorrelation function of a random signal plus its delayed and attenuated versions

From this equation it is concluded that indeed the term with a delay of T has been removed. One may argue that, instead, the term $\alpha^2 X(t-2T)$ has been introduced. That is right, but it is not unreasonable to assume that the reflection coefficient α is (much) less than unity, so that this newly introduced term is smaller by the factor of α compared to the distortion in the received signal. If this is nevertheless unacceptable then a further reduction is achieved by also adding the term $\alpha^2 Y(t-2T)$ to the received signal. This removes the distortion at $2T$ and in its turn introduces a term that is still smaller by an amount of α^3 at a delay of $3T$, etc. In this way the distortion may be reduced to an arbitrary small amount.

□

Apart from these examples there are several applications that use correlation as the basic signal processing method for extracting information from an observation.

2.2.4 Measuring Correlation Functions

In a practical situation it is impossible to measure a correlation function. This is due to the fact that we will never have available the entire ensemble of sample functions of the process in question. Even if we did have them then it would nevertheless be impossible to cope with an infinite number of sample functions. Thus we have to confine ourselves to a limited class of processes, e.g. to the class of ergodic processes. We have established before that most of the time it is difficult or even impossible to determine whether a process is ergodic or not. Unless the opposite is clear, we will assume ergodicity in practice; this greatly simplifies matters, especially for measuring correlation functions. This assumption enables the wanted correlation function based on just a single sample function to be determined, as is evident from Equation (2.25).

In Figure 2.7 a block schematic is shown for a possible set-up to measure a cross-correlation function $R_{XY}(\tau)$, where the assumption has to be made that the processes $X(t)$ and $Y(t)$ are jointly ergodic. The sample functions $x(t)$ and $y(t)$ should be applied to the inputs at least starting at $t=-T+\tau$ up until $t=T+\tau$. The signal $x(t)$ is delayed and applied to a multiplier whereas $y(t)$ is applied undelayed to a second input of the same multiplier. The multiplier's output is applied to an integrator that integrates over a period $2T$. Looking at this scheme we conclude that the measured output is

$$R_o(\tau,T)=\frac{1}{2T}\int_{-T+\tau}^{T+\tau}x(t-\tau)\,y(t)\,\mathrm{d}t=\frac{1}{2T}\int_{-T}^{T}x(t)\,y(t+\tau)\,\mathrm{d}t \tag{2.61}$$

If the integration time $2T$ is taken long enough, and remembering the assumption on the jointly ergodicity of $X(t)$ and $Y(t)$, the measured value $R_o(\tau,T)$ will approximate $R_{XY}(\tau)$. By

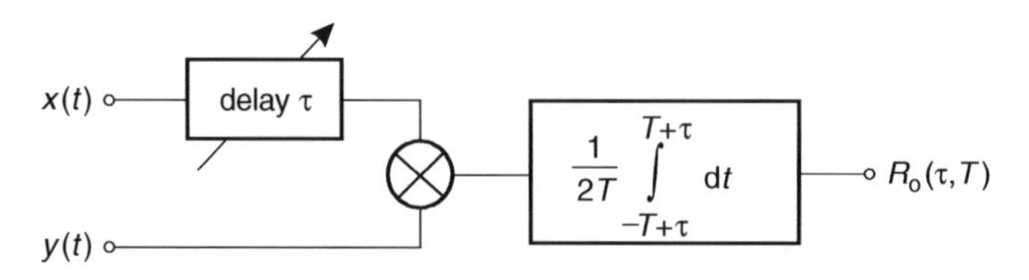

Figure 2.7 Measurement scheme for correlation functions

varying τ the function can be measured for different values of the argument. By simply short-circuiting the two inputs and applying a single signal $x(t)$ to this common input, the autocorrelation function $R_{XX}(\tau)$ is measured.

In practice only finite measuring times can be realized. In general this will introduce an error in the measured result. In the next example this point will be further elaborated.

Example 2.6:

Let us consider the example that has been subject of our studies several times before, namely the cosine waveform with amplitude A and random phase that has a uniform distribution over one period of the cosine. This process has been described in Example 2.1. Suppose we want to measure the autocorrelation function of this process using the set-up given in Figure 2.7. The inputs are short-circuited and the signal is applied to these common inputs. If the given process is substituted in Equation (2.61), we find

$$\begin{aligned} R_o(\tau, T) &= \frac{1}{2T} A^2 \int_{-T}^{T} \cos(\omega t - \theta) \cos(\omega t + \omega\tau - \theta)\, dt \\ &= \frac{A^2}{4T} \int_{-T}^{T} [\cos \omega\tau + \cos(2\omega t + \omega\tau - 2\theta)]\, dt \end{aligned} \tag{2.62}$$

In this equation the random variable Θ has not been used, but the specific value θ that corresponds to the selected realization of the process. The first term in the integrand of this integral produces $(A^2/2)\cos(\omega\tau)$, the value of the autocorrelation function of this process, as was concluded in Example 2.1. The second term in the integrand must be a measurement error. The magnitude of this error is determined by evaluating the corresponding integral. This yields

$$e(\tau, T) = \frac{A^2}{2} \cos(\omega\tau - 2\theta) \frac{\sin(2\omega T)}{2\omega T} \tag{2.63}$$

The error has an oscillating character as a function of T, while the absolute value of the error decreases inversely with T. At large values of T the error approaches 0. If, for example, the autocorrelation function has to be measured with an accuracy of 1%, then the condition $1/(2\omega T) < 0.01$ should be fulfilled, or equivalently the measurement time should satisfy $2T > 100/\omega$.

Although this analysis looks nice, its applicability is limited. In practice the autocorrelation function is not known beforehand; that is why we want to measure it. Thus the above error analysis cannot be carried out. The solution to this problem consists of doing a best-effort measurement and then to make an estimate of the error in the correlation function. Looking back, it possible to decide whether the measurement time was long enough for the required accuracy. If not, the measurement can be redone using a larger (estimated) measurement time based on the error analysis. In this way accuracy can be iteratively improved.

□

2.2.5 Covariance Functions

The concept of covariance of two random variables can be extended to stochastic processes. The autocovariance function of a stochastic process is defined as

$$C_{XX}(t, t+\tau) \triangleq \mathrm{E}[\{X(t) - \mathrm{E}[X(t)]\}\{X(t+\tau) - \mathrm{E}[X(t+\tau)]\}] \tag{2.64}$$

This can be written as

$$C_{XX}(t, t+\tau) = R_{XX}(t, t+\tau) - \mathrm{E}[X(t)]\,\mathrm{E}[X(t+\tau)] \tag{2.65}$$

The cross-covariance function of two processes $X(t)$ and $Y(t)$ is defined as

$$C_{XY}(t, t+\tau) \triangleq \mathrm{E}[\{X(t) - \mathrm{E}[X(t)]\}\{Y(t+\tau) - \mathrm{E}[Y(t+\tau)]\}] \tag{2.66}$$

or

$$C_{XY}(t, t+\tau) = R_{XY}(t, t+\tau) - \mathrm{E}[X(t)]\,\mathrm{E}[Y(t+\tau)] \tag{2.67}$$

For processes that are at least jointly wide-sense stationary the second expressions in the right-hand sides of Equations (2.65) and (2.67) can be simplified, yielding

$$C_{XX}(\tau) = R_{XX}(\tau) - \overline{X}^2 \tag{2.68}$$

and

$$C_{XY}(\tau) = R_{XY}(\tau) - \overline{X}\,\overline{Y} \tag{2.69}$$

respectively. From Equation (2.68) and property 4 of the autocorrelation function in Section 2.2.1 it follows immediately that

$$\lim_{|\tau|\to\infty} C_{XX}(\tau) = 0 \tag{2.70}$$

provided the process $X(t)$ does not have a periodic component. If in Equation (2.64) the value $\tau = 0$ is used we obtain the variance of the process. In the case of wide-sense stationary processes the variance is independent of time, and using Equation (2.68) we arrive at

$$\sigma_X^2 \triangleq \mathrm{E}[\{X(t) - \mathrm{E}[X(t)]\}^2] = C_{XX}(0) = R_{XX}(0) - \overline{X}^2 \tag{2.71}$$

If for two processes

$$C_{XY}(t, t+\tau) \equiv 0 \tag{2.72}$$

then these processes are called uncorrelated processes. According to Equation (2.67) this has as a consequence

$$R_{XY}(t, t+\tau) = \mathrm{E}[X(t)]\,\mathrm{E}[Y(t+\tau)] \tag{2.73}$$

Since this latter equation is identical to Equation (2.55), it follows that independent processes are uncorrelated. The converse is not necessarily true, unless the processes are jointly Gaussian processes (see Section 2.3).

2.2.6 Physical Interpretation of Process Parameters

In the previous sections stochastic processes have been described from a mathematical point of view. In practice we want to relate these descriptions to physical concepts such as a signal, represented, for example, by a voltage or a current. In these cases the following physical interpretations are connected to the parameters of the stochastic processes:

- The mean $\overline{X(t)}$ is proportional to the d.c. component of the signal.
- The squared mean value $\overline{X(t)}^2$ is proportional to the power in the d.c. component of the signal.
- The mean squared value $\overline{X^2(t)}$ is proportional to the total average power of the signal.
- The variance $\sigma_X^2 \triangleq \overline{X^2(t)} - \overline{X(t)}^2$ is proportional to the power in the time-varying components of the signal, i.e. the a.c. power.
- The standard deviation σ_X is the square root of the mean squared value of the time-varying components, i.e. the root-mean-square (r.m.s.) value.

In Chapter 6 the proportionality factors will be deduced. Now it suffices to say that this proportionality factor becomes unity in case the load is purely resistive and equal to one.

Although the above interpretations serve to make the engineer familiar with the practical value of stochastic processes, it must be stressed that they only apply to the special case of signals that can be modelled as ergodic processes.

2.3 GAUSSIAN PROCESSES

Several processes can be modelled by what is called a Gaussian process; among these is the thermal noise process that will be presented in Chapter 6. As the name suggests, these processes are described by Gaussian distributions. Recall that the probability density function of a Gaussian random variable X is defined by [1–5]

$$f_X(x) = \frac{1}{\sigma_X\sqrt{2\pi}}\exp\left[-\frac{(x-\overline{X})^2}{2\sigma_X^2}\right] \tag{2.74}$$

The Gaussian distribution is frequently encountered in engineering and science. When considering two jointly Gaussian random variables X and Y we sometimes need the joint probability density function, as will become apparent in the sequel

$$f_{XY}(x,y) = \frac{1}{2\pi\sigma_X\sigma_Y\sqrt{1-\rho^2}}\exp\left\{\frac{-1}{2(1-\rho^2)}\left[\frac{(x-\overline{X})^2}{\sigma_X^2} - \frac{2\rho(x-\overline{X})(y-\overline{Y})}{\sigma_X\sigma_Y} + \frac{(y-\overline{Y})^2}{\sigma_Y^2}\right]\right\} \tag{2.75}$$

where ρ is the correlation coefficient defined by

$$\rho \triangleq \frac{\mathrm{E}[(X-\overline{X})(Y-\overline{Y})]}{\sigma_X\sigma_Y} \tag{2.76}$$

For N jointly Gaussian random variables $X_1, X_2, \ldots, X_N$, the joint probability density function reads

$$f_{X_1X_2\cdots X_N}(x_1, x_2, \ldots, x_N) = \frac{|\mathbf{C}_X^{-1}|^{\frac{1}{2}}}{(2\pi)^{N/2}}\exp\left[-\frac{(\mathbf{x}-\overline{\mathbf{X}})^{\mathrm{T}}\mathbf{C}_X^{-1}(\mathbf{x}-\overline{\mathbf{X}})}{2}\right] \tag{2.77}$$

where we define the vector

$$\mathbf{x}-\overline{\mathbf{X}} \triangleq \begin{bmatrix} x_1-\overline{X_1} \\ x_2-\overline{X_2} \\ \vdots \\ x_N-\overline{X_N} \end{bmatrix} \tag{2.78}$$

and the covariance matrix

$$\mathbf{C}_X \triangleq \begin{bmatrix} C_{11} & C_{12} & \cdots & C_{1N} \\ C_{21} & C_{22} & \cdots & C_{2N} \\ \vdots & \vdots & & \vdots \\ C_{N1} & C_{N2} & \cdots & C_{NN} \end{bmatrix} \tag{2.79}$$

In the foregoing we used $\mathbf{x}^{\mathrm{T}}$ for the matrix transpose, $\mathbf{C}^{-1}$ for the matrix inverse and $|\mathbf{C}|$ for the determinant. The elements of the covariance matrix are defined by

$$C_{ij} \triangleq \mathrm{E}(X_i-\overline{X_i})(X_j-\overline{X_j})] \tag{2.80}$$

The diagonal elements of the covariance matrix equal the variances of the various random variables, i.e. $C_{ii} = \sigma_{X_i}^2$. It is easily verified that Equations (2.74) and (2.75) are special cases of Equation (2.77).

Gaussian variables as described above have a few interesting properties, which have their consequences for Gaussian processes. These properties are [1–5]:

1. Gaussian random variables are completely specified only by their first and second order moments, i.e. by their means, variances and covariances. This is immediately apparent, since these are the only quantities present in Equation (2.77).
2. When Gaussian random variables are uncorrelated, they are independent. For uncorrelated random variables (i.e. $\rho = 0$) the covariance matrix is reduced to a diagonal matrix. It is easily verified from Equation (2.77) that in such a case the probability density function of N variables can be written as the product of N functions of the type given in Equation (2.74).
3. A linear combination of Gaussian random variables produces another Gaussian variable. For the proof of this see reference [2] and Problem 8.3.

We are now able to define a Gaussian stochastic process. Referring to Equation (2.77), a process $X(t)$ is called a Gaussian process if the random variables $X_1 = X(t_1)$, $X_2 = X(t_2), \ldots, X_N = X(t_N)$ are jointly Gaussian and thus satisfy

$$f_X(x_1, \ldots, x_N; t_1, \ldots, t_N) = \frac{|\mathbf{C}_X^{-1}|^{\frac{1}{2}}}{(2\pi)^{N/2}} \exp\left[-\frac{(\mathbf{x}-\overline{\mathbf{X}})^{\mathrm{T}}\mathbf{C}_X^{-1}(\mathbf{x}-\overline{\mathbf{X}})}{2}\right] \tag{2.81}$$

for all arbitrary N and for any set of times $t_1, \ldots, t_N$. Now the mean values $\overline{X_i}$ of $X(t_i)$ are

$$\overline{X_i} = \mathrm{E}[X(t_i)] \tag{2.82}$$

and the elements of the covariance matrix are

$$\begin{aligned} C_{ij} &= \mathrm{E}[(X_i - \overline{X_i})(X_j - \overline{X_j})] \\ &= \mathrm{E}[\{X(t_i) - \mathrm{E}[X(t_i)]\}\{X(t_j) - \mathrm{E}[X(t_j)]\}] \\ &= C_{XX}(t_i, t_j) \end{aligned} \tag{2.83}$$

which is the autocovariance function as defined by Equation (2.64).

Gaussian processes have a few interesting properties.

Properties of Gaussian Processes

1. Gaussian processes are completely specified by their mean $\mathrm{E}[X(t)]$ and autocorrelation function $R_{XX}(t_i, t_j)$.
2. A wide-sense stationary Gaussian process is also strict-sense stationary.
3. If the jointly Gaussian processes $X(t)$ and $Y(t)$ are uncorrelated, then they are independent.
4. If the Gaussian process $X(t)$ is passed through a linear time-invariant system, then the corresponding output process $Y(t)$ is also a Gaussian process.

These properties are closely related to the properties of jointly Gaussian random variables previously discussed in this section. Let us briefly comment on the properties:

1. We saw before that the joint probability density function is completely determined when the mean and autocovariance are known. However, these two quantities as functions of time in their turn determine the autocorrelation function (see Equation (2.68)).
2. The nth-order probability density function of a Gaussian process only depends on the two functions $\mathrm{E}[X(t)]$ and $C_{XX}(t, t+\tau)$. When the process is wide-sense stationary then these functions do not depend on the absolute time t, and as a consequence

$$f_X(x_1, \ldots, x_N; t_1, \ldots, t_N) = f_X(x_1, \ldots, x_N; t_1 + \tau, \ldots, t_N + \tau) \tag{2.84}$$

   Since this is valid for all arbitrary N and all τ, it is concluded that the process is strict-sense stationary.
3. This property is a straightforward consequence of the property of jointly random variables discussed before.
4. Passing a process through a linear time-invariant system is described by a convolution, which may be considered as the limit of a weighted sum of samples of the input process. From the preceding we know that a linear combination of Gaussian variables produces another Gaussian variable.

2.4 COMPLEX PROCESSES

A complex stochastic process is defined by

$$Z(t) \triangleq X(t) + \mathrm{j}Y(t) \tag{2.85}$$

with $X(t)$ and $Y(t)$ real stochastic processes. Such a process is said to be stationary if $X(t)$ and $Y(t)$ are jointly stationary. Expectation and the autocorrelation function of a complex stochastic process are defined as

$$\mathrm{E}[Z(t)] \triangleq \mathrm{E}[X(t) + \mathrm{j}Y(t)] = \mathrm{E}[X(t)] + \mathrm{j}\mathrm{E}[Y(t)] \tag{2.86}$$

and

$$R_{ZZ}(t, t+\tau) \triangleq \mathrm{E}[Z^*(t)\, Z(t+\tau)] \tag{2.87}$$

where * indicates the complex conjugate.

For the autocovariance function the definition of Equation (2.87) is used, where $Z(t)$ is replaced by the stochastic process $Z(t) - \mathrm{E}[Z(t)]$. This yields

$$C_{ZZ}(t, t+\tau) = R_{ZZ}(t, t+\tau) - \mathrm{E}^*[Z(t)]\, \mathrm{E}[Z(t+\tau)] \tag{2.88}$$

The cross-correlation function of two complex processes $Z_i(t)$ and $Z_j(t)$ reads

$$R_{Z_iZ_j}(t, t+\tau) = \mathrm{E}[Z_i^*(t)\, Z_j(t+\tau)] \tag{2.89}$$

and the cross-covariance function is found from Equation (2.89) by replacing $Z_{i,j}(t)$ with $Z_{i,j}(t) - \mathrm{E}[Z_{i,j}(t)]$; this yields

$$C_{Z_iZ_j}(t, t+\tau) = R_{Z_iZ_j}(t, t+\tau) - \mathrm{E}^*[Z_i(t)]\,\mathrm{E}[Z_j(t+\tau)] \tag{2.90}$$

In the chapters that follow we will work exclusively with real processes, unless it is explicitly indicated that complex processes are considered.

One may wonder why the correlation functions of complex processes are defined in the way it has been done in Equations (2.87) and (2.89). The explanation for this arises from an engineering point of view; namely the given expressions of the correlation functions evaluated for $\tau = 0$ have to result in the expectation of the squared process for real processes. In engineering calculations real processes are replaced many times by complex processes of the form $I = \hat{I}\exp(\mathrm{j}\omega t)$ (for a current) or $V = \hat{V}\exp(\mathrm{j}\omega t)$ (for a voltage). In these cases the correlation function for $\tau = 0$ should be a quantity that is proportional to the mean power. The given definitions satisfy this requirement.

2.5 DISCRETE-TIME PROCESSES

In Chapter 1 the discrete-time process was introduced by sampling a continuous stochastic process. However, at this point we are not yet able to develop a sampling theorem for stochastic processes analogously to that for deterministic signals [1]. We will derive such a theorem in Chapter 3. This means that in this section we deal with random sequences as such, irrespective of their origin. In Chapter 1 we introduced the notation $X[n]$ for random sequences. In this section we will assume that the sequences are real. However, they can be complex valued. Extension to complex discrete-time processes is similar to what was derived in the former section.

In the next subsection we will resume the most important properties of discrete-time processes. Since such processes are actually special cases of continuous stochastic processes the properties are self-evident.

2.5.1 Mean, Correlation Functions and Covariance Functions

The mean value of a discrete-time process is found by

$$\mathrm{E}[X[n]] = \overline{X[n]} \triangleq \int_{-\infty}^{\infty} x\, f_X(x; n)\,\mathrm{d}x \tag{2.91}$$

Recall that the process is time-discrete but the x values are continuous, so that indeed the expectation (or ensemble mean) is written as an integral over a continuous probability density function. This function describes the random variable $X[n]$ by considering all

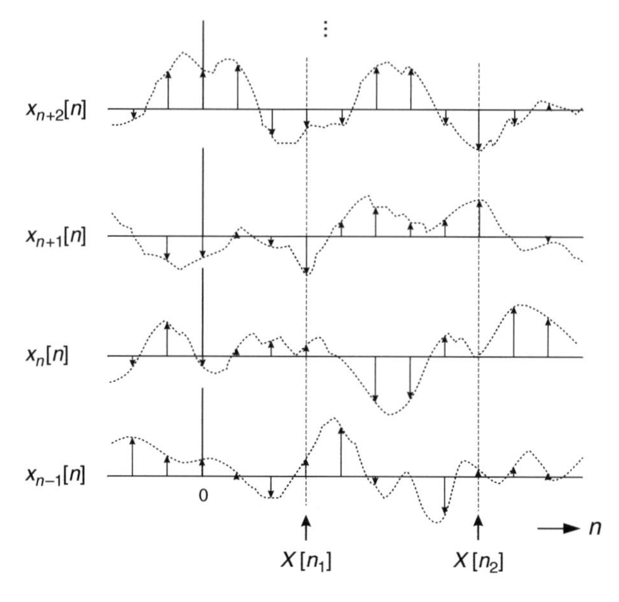

Figure 2.8 Random variables $X[n_1]$ and $X[n_2]$ that arise when considering the ensemble values of the discrete-time process $X[n]$ at fixed positions n_1 and n_2

possible ensemble realizations of the process at a fixed integer position for example n_1 (see Figure 2.8). For real processes the autocorrelation sequence is defined as

$$R_{XX}[n_1, n_2] \triangleq \mathrm{E}[X[n_1]\,X[n_2]] \triangleq \iint x_1 x_2\, f_X(x_1, x_2; n_1, n_2)\,\mathrm{d}x_1\,\mathrm{d}x_2 \tag{2.92}$$

where the process is now considered at two positions n_1 and n_2 jointly (see again Figure 2.8).

For the autocovariance sequence of this process (compare to Equation (2.65))

$$C_{XX}[n_1, n_2] = R_{XX}[n_1, n_2] - \mathrm{E}[X[n_1]]\,\mathrm{E}[X[n_2]] \tag{2.93}$$

The cross-correlation and cross-covariance sequences are defined analogously, namely respectively as

$$R_{XY}[n, n+m] \triangleq \mathrm{E}[X[n]\,Y[n+m]] \tag{2.94}$$

and (compare with Equation (2.67))

$$C_{XY}[n, n+m] \triangleq R_{XY}[n, n+m] - \mathrm{E}[X[n]]\,\mathrm{E}[Y[n+m]] \tag{2.95}$$

A discrete-time process is called wide-sense stationary if the next two conditions hold jointly:

$$\mathrm{E}[X[n]] = \text{constant} \tag{2.96}$$

$$R_{XX}[n, n+m] = R_{XX}[m] \tag{2.97}$$

i.e. the autocorrelation sequence only depends on the difference m of the integer positions.

Two discrete-time processes are jointly wide-sense stationary if they are individually wide-sense stationary and moreover

$$R_{XY}[n, n+m] = R_{XY}[m] \tag{2.98}$$

i.e. the cross-correlation sequence only depends on the difference m of the integer positions.

The time average of a discrete-time process is defined as

$$\mathrm{A}[X[n]] \triangleq \lim_{N\to\infty} \frac{1}{2N+1} \sum_{n=-N}^{N} X[n] \tag{2.99}$$

A wide-sense stationary discrete-time process is ergodic if the two conditions are satisfied

$$\mathrm{A}[X[n]] = \mathrm{E}[X[n]] = \overline{X[n]} \tag{2.100}$$

and

$$\mathrm{A}[X[n]\,X[n+m]] = \mathrm{E}[X[n]\,X[n+m]] = R_{XX}[m] \tag{2.101}$$

2.6 SUMMARY

An ensemble is the set of all possible realizations of a stochastic process $X(t)$. A realization or sample function is provided by a random selection out of this ensemble. For the description of stochastic processes a parameter is added to the well-known definitions of the probability distribution function and the probability density function, namely the time parameter. This means that these functions in the case of a stochastic process are as a rule functions of time. When considering stationary processes certain time dependencies disappear; we thus arrive at first-order and second-order stationary processes, which are useful for practical applications.

The correlation concept is in random signal theory, analogously to probability theory, defined as the expectation of the product of two random variables. For the autocorrelation function these variables are $X(t)$ and $X(t+\tau)$, while for the cross-correlation function of two processes the quantities $X(t)$ and $Y(t+\tau)$ are used in the definition.

A wide-sense stationary process is a process where the mean value is constant and the autocorrelation function only depends on τ, not on the absolute time t. When calculating the expectations the time t is considered as a parameter; i.e. in these calculations t is given a fixed value. The random variable is the variable based on which outcome of the realization is chosen from the ensemble. When talking about 'mean' we have in mind the ensemble mean, unless it is explicitly indicated that a different definition is used (for instance the time average). In the case of an ergodic process the first- and second-order time averages equal the first- and second-order ensemble means, respectively. The theorem that has been presented on cyclo-stationary processes plays an important role in 'making stationary' certain classes of processes. The covariance functions of stochastic processes are the correlation functions of these processes minus their own process mean values.

Physical interpretations of several stochastic concepts have been presented. Gaussian processes get special attention as they are of practical importance and possess a few interesting and convenient properties. Complex processes are defined analogously to the usual method for complex variables.

Finally, several definitions and properties of continuous stochastic processes are redefined for discrete-time processes.

2.7 PROBLEMS

2.1 All sample functions of a stochastic process are constant, i.e. $X(t) = C =$ constant, where C is a discrete random variable that may assume the values $C_1 = 1$, $C_2 = 3$ and $C_3 = 4$, with probabilities of 0.5, 0.3 and 0.2, respectively.

(a) Determine the probability density function of $X(t)$.

(b) Calculate the mean and variance of $X(t)$.

2.2 Consider a stationary Gaussian process with a mean of zero.

(a) Determine and sketch the probability density function of this process after passing it through an ideal half-wave rectifier.

(b) Same question for the situation where the process is applied to a full-wave rectifier.

2.3 A stochastic process comprises four sample functions, namely $x(t, s_1) = 1$, $x(t, s_2) = t$, $x(t, s_3) = \cos t$ and $x(t, s_4) = 2 \sin t$, which occur with equal probabilities.

(a) Determine the probability density function of $X(t)$.

(b) Is the process stationary in any sense?

2.4 Consider the process

$$X(t) = \sum_{n=1}^{N} A_n \cos(\omega_n t) + B_n \sin(\omega_n t)$$

where A_n and B_n are random variables that are mutually uncorrelated, have zero mean and of which

$$\mathrm{E}[A_n^2] = \mathrm{E}[B_n^2] = \sigma^2$$

The quantities $\{\omega_n\}$ are constants.

(a) Calculate the autocorrelation function of $X(t)$.

(b) Is the process wide-sense stationary?

2.5 Consider the stochastic process $X(t) = A\cos(\omega_0 t) + B\sin(\omega_0 t)$, with ω_0 a constant and A and B random variables. What are the conditions for A and B in order for $X(t)$ to be wide-sense stationary?

2.6 Consider the process $X(t) = A\ \cos(\omega_0 t - \Theta)$, where A and Θ are independent random variables and Θ is uniformly distributed on the interval $(0, 2\pi]$.

(a) Is this process wide-sense stationary?

(b) Is it ergodic?

2.7 Consider the two processes

$$X(t) = A\cos(\omega_0 t) + B\sin(\omega_0 t)$$
$$Y(t) = A\cos(\omega_0 t) - B\sin(\omega_0 t)$$

with A and B independent random variables, both with zero mean and equal variance of σ^2. The angular frequency ω_0 is constant.

(a) Are the processes $X(t)$ and $Y(t)$ wide-sense stationary?

(b) Are they jointly wide-sense stationary?

2.8 Consider the stochastic process $X(t) = A\sin(\omega_0 t - \Theta)$, with A and ω_0 constants, and Θ a random variable that is uniformly distributed on the interval $(0, 2\pi]$. We define a new process by means of $Y(t) = X^2(t)$.

(a) Are $X(t)$ and $Y(t)$ wide-sense stationary?

(b) Calculate the autocorrelation function of $Y(t)$.

(c) Calculate the cross-correlation function of $X(t)$ and $Y(t)$.

(d) Are $X(t)$ and $Y(t)$ jointly wide-sense stationary?

(e) Calculate and sketch the probability distribution function of $Y(t)$.

(f) Calculate and sketch the probability density function of $Y(t)$.

2.9 Repeat Problem 2.8 when $X(t)$ is half-wave rectified. Use Matlab to plot the autocorrelation function.

2.10 Repeat Problem 2.8 when $X(t)$ is full-wave rectified. Use Matlab to plot the autocorrelation function.

2.11 The function $p(t)$ is defined as

$$p(t) = \begin{cases} 1, & 0 \le t \le \frac{3}{4}T \\ 0, & \text{all other values of } t \end{cases}$$

By means of this function we define the stochastic process

$$X(t) = \sum_{n=-\infty}^{\infty} p(t - nT - \Theta)$$

where Θ is a random variable that is uniformly distributed on the interval $[0, T)$.

(a) Sketch a possible realization of $X(t)$.

(b) Calculate the mean value of $X(t)$.

(c) Calculate and sketch the autocorrelation function of $X(t)$.

(d) Calculate and sketch the probability distribution function of $X(t)$.

(e) Calculate and sketch the probability density function of $X(t)$.

(f) Calculate the variance of $X(t)$.

2.12 Two functions $p_1(t)$ and $p_2(t)$ are defined as

$$p_1(t) = \begin{cases} 1, & 0 \le t \le \frac{1}{3}T \\ 0, & \text{all other values of } t \end{cases}$$

and

$$p_2(t) = \begin{cases} 1, & 0 \le t \le \frac{2}{3}T \\ 0, & \text{all other values of } t \end{cases}$$

Based on these functions the stochastic processes $X(t)$ and $Y(t)$ are defined as

$$X(t) = \sum_{n=-\infty}^{\infty} p_1(t - nT - \Theta)$$

$$Y(t) = \sum_{n=-\infty}^{\infty} p_2(t - nT - \Theta)$$

and

$$W(t) \triangleq X(t) + Y(t)$$

where Θ is a random variable that is uniformly distributed on the interval $[0, T)$.

(a) Sketch possible realizations of $X(t)$ and $Y(t)$.
(b) Calculate and sketch the autocorrelation function of $X(t)$.
(c) Calculate and sketch the autocorrelation function of $Y(t)$.
(d) Calculate and sketch the autocorrelation function of $W(t)$.
(e) Calculate the power in $W(t)$, i.e. $\mathrm{E}[W^2(t)]$.

2.13 The processes $X(t)$ and $Y(t)$ are independent with a mean value of zero and autocorrelation functions $R_{XX}(\tau) = \exp(-|\tau|)$ and $R_{YY}(\tau) = \cos(2\pi\tau)$, respectively.

(a) Derive the autocorrelation function of the sum $W_1(t) = X(t) + Y(t)$.
(b) Derive the autocorrelation function of the difference $W_2(t) = X(t) - Y(t)$.
(c) Calculate the cross-correlation function of $W_1(t)$ and $W_2(t)$.

2.14 In Figure 2.9 the autocorrelation function of a wide-sense stationary stochastic process $X(t)$ is given.

(a) Calculate the value of $\mathrm{E}[X(t)]$.
(b) Calculate the value of $\mathrm{E}[X^2(t)]$.
(c) Calculate the value of σ_X^2.

2.15 Starting from the wide-sense stationary process $X(t)$ we define a new process as $Y(t) = X(t) - X(t+T)$.

(a) Show that the mean value of $Y(t)$ is zero, even if the mean value of $X(t)$ is not zero.
(b) Show that $\sigma_Y^2 = 2\{R_{XX}(0) - R_{XX}(T)\}$.

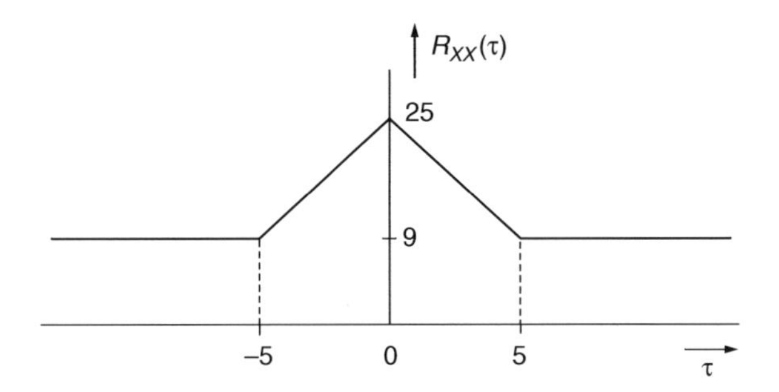

Figure 2.9

(c) If $Y(t) = X(t) + X(t + T)$ find expressions for $E[Y(t)]$ and σ_Y^2. Compare these results with the answers found in (a) and (b).

2.16 Determine for each of the following functions whether it can be the autocorrelation function of a real wide-sense stationary process $X(t)$.

(a) $R_{XX}(\tau) = \mathrm{u}(\tau)\exp(-\tau)$.
(b) $R_{XX}(\tau) = 3\sin(7\tau)$.
(c) $R_{XX}(\tau) = (1 + \tau^2)^{-1}$.
(d) $R_{XX}(\tau) = -\cos(2\tau)\exp(-|\tau|)$.
(e) $R_{XX}(\tau) = 3[\sin(4\tau)/(4\tau)]^2$.
(f) $R_{XX}(\tau) = 1 + 3\sin(8\tau)/(8\tau)$.

2.17 Consider the two processes $X(t)$ and $Y(t)$. Find expressions for the autocorrelation function of $W(t) = X(t) + Y(t)$ in the case where:

(a) $X(t)$ and $Y(t)$ are correlated;
(b) $X(t)$ and $Y(t)$ are uncorrelated;
(c) $X(t)$ and $Y(t)$ are uncorrelated and have mean values of zero.

2.18 The voltage of the output of a noise generator is measured using a d.c. voltmeter and a true root-mean-square (r.m.s.) meter that has a series capacitor at its input. The noise is known to be Gaussian and stationary. The reading of the d.c. meter is 3 V and that of the r.m.s. meter is 2 V. Derive an expression for the probability density function of the noise and make a plot of it using Matlab.

2.19 Two real jointly wide-sense stationary processes $X(t)$ and $Y(t)$ are used to define two complex processes as follows:

$$Z_1(t) = X(t) + \mathrm{j}Y(t)$$

and

$$Z_2(t) = X(t - T) - \mathrm{j}Y(t - T)$$

Calculate the cross-correlation function of the processes $Z_1(t)$ and $Z_2(t)$.

2.20 A voltage source is described as $V = 5\cos(\omega_0 t - \Theta)$, where Θ is a random variable that is uniformly distributed on $[0, 2\pi)$. This source is applied to an electric circuit and as a consequence the current flowing through the circuit is given by $I = 2\cos(\omega_0 t - \Theta + \pi/6)$.

(a) Calculate the cross-correlation function of V and I.

(b) Calculate the electrical power that is absorbed by the circuit.

(c) If in general an harmonic voltage at the terminals of a circuit is described by its complex notation $V = \hat{V}\exp[j(\omega t - \Theta)]$ and the corresponding current that is flowing into the circuit by a similar notation $I = \hat{I}\exp[j(\omega t - \Theta + \phi)]$, with ϕ a constant, show that the electrical power absorbed by the circuit is written as $P_{\text{el}} = (\hat{V}\hat{I}\cos\phi)/2$.

2.21 Consider a discrete-time wide-sense stationary process $X[n]$. Show that for such a process

$$3R_{XX}[0] \geq |4R_{XX}[1] + 2R_{XX}[2]|$$

3

Spectra of Stochastic Processes

In Chapter 2 stochastic processes have been considered in the time domain exclusively; i.e. we used such concepts as the autocorrelation function, the cross-correlation function and the covariance function to describe the processes. When dealing with deterministic signals, we have the frequency domain at our disposal as a means to an alternative, dual description. One may wonder whether for stochastic processes a similar duality exists. This question is answered in the affirmative, but the relationship between time domain and frequency domain descriptions is different compared to deterministic signals. Hopping from one domain to the other is facilitated by the well-known Fourier transform and its inverse transform. A complicating factor is that for a random waveform (a sample function of the stochastic process) the Fourier transform generally does not exist.

3.1 THE POWER SPECTRUM

Due to the problems with the Fourier transform, a theoretical description of stochastic processes must basically start in the time domain, as given in Chapter 2. In this chapter we will confine ourselves exclusively to wide-sense stationary processes with the autocorrelation function $R_{XX}(\tau)$. Let us assume that it is allowed to apply the Fourier transform to $R_{XX}(\tau)$.

Theorem 3

The Wiener–Khinchin relations are

$$S_{XX}(\omega) = \int_{-\infty}^{\infty} R_{XX}(\tau) \exp(-j\omega\tau)\, d\tau \tag{3.1}$$

$$R_{XX}(\tau) = \frac{1}{2\pi} \int_{-\infty}^{\infty} S_{XX}(\omega) \exp(j\omega\tau)\, d\omega \tag{3.2}$$

Introduction to Random Signals and Noise W. van Etten

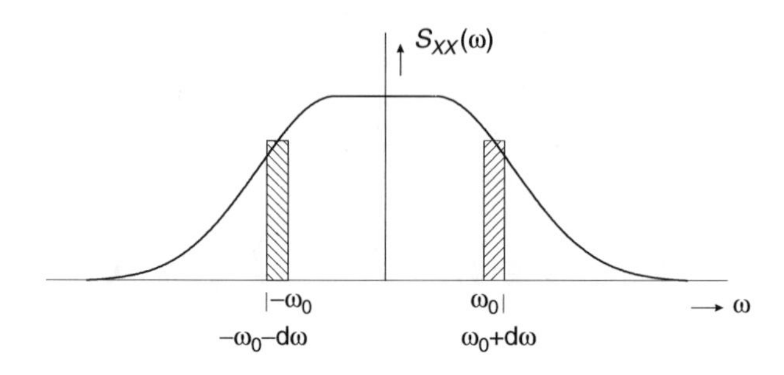

Figure 3.1 Interpretation of $S_{XX}(\omega)$

The function $S_{XX}(\omega)$ has an interesting interpretation, as will follow from the sequel. For that purpose we put the variable τ equal to zero in Equation (3.2). This yields

$$R_{XX}(0) = \mathrm{E}[X^2(t)] = \frac{1}{2\pi}\int_{-\infty}^{\infty} S_{XX}(\omega)\,\mathrm{d}\omega \tag{3.3}$$

However, from Equation (2.18) it is concluded that $R_{XX}(0)$ equals the mean squared value of the process; this is called the mean power of the process, or just the power of the process. Now it follows from Equation (3.3) that $S_{XX}(\omega)$ represents the way in which the total power of the process is spread over the different frequency components. This is clear since integrating $S_{XX}(\omega)$ over the entire frequency axis produces the total power of the process. In other words, $2S_{XX}(\omega_0)\mathrm{d}\omega/(2\pi)$ is the power at the output of the bandpass filter with the passband transfer function

$$H(\omega) = \begin{cases} 1, & \omega_0 < |\omega| < \omega_0 + \mathrm{d}\omega \\ 0, & \text{elsewhere} \end{cases} \tag{3.4}$$

when the input of this filter consists of the process $X(t)$. This is further explained by Figure 3.1. Due to this interpretation the function $S_{XX}(\omega)$ is called the power spectral density, or briefly the power spectrum of the process $X(t)$. The Wiener–Khinchin relations state that the autocorrelation function and the power spectrum of a wide-sense stationary process are a Fourier transform pair. From the given interpretation the properties of the Fourier transform are as follows.

Properties of $S_{XX}(\omega)$

1. $S_{XX}(\omega) \geq 0$ (3.5)
2. $S_{XX}(-\omega) = S_{XX}(\omega)$, for a real process $X(t)$ (3.6)
3. $\mathrm{Im}\{S_{XX}(\omega)\} \equiv 0$ (3.7)
   where $\mathrm{Im}\{\cdot\}$ is defined as the imaginary part of the quantity between the braces
4. $\frac{1}{2\pi}\int_{-\infty}^{\infty} S_{XX}(\omega)\,\mathrm{d}\omega = \mathrm{E}[X^2(t)] = R_{XX}(0) = P_{XX}$ (3.8)

Proofs of the properties:

1. Property 1 is connected to the interpretation of $S_{XX}(\omega)$ and a detailed proof will be given in Chapter 4.

2. Property 2 states that the power spectrum is an even function of ω. The proof of this property is based on Fourier theory and the fact that for a real process the autocorrelation function $R_{XX}(\tau)$ is real and even. The proof proceeds as follows:

$$S_{XX}(\omega) = \int_{-\infty}^{\infty} R_{XX}(\tau)[\cos(\omega\tau) - \mathrm{j}\sin(\omega\tau)]\,\mathrm{d}\tau \tag{3.9}$$

   Since $R_{XX}(\tau)$ is real and even, the product of this function and a sine is odd. Therefore, this product makes no contribution to the integral, which runs over a symmetrical range of the integration variable. The remaining part is a product of $R_{XX}(\tau)$ and a cosine, both being even, resulting in an even function of ω.

3. The third property, $S_{XX}(\omega)$ being real, is proved as follows. Let us define the complex process $X(t) = R(t) + \mathrm{j}I(t)$, where $R(t)$ and $I(t)$ are real processes and represent the real and imaginary part of $X(t)$, respectively. Then after some straightforward calculations the autocorrelation function of $X(t)$ is

$$R_{XX}(\tau) = R_{RR}(\tau) + R_{II}(\tau) + \mathrm{j}[R_{RI}(\tau) - R_{IR}(\tau)] \tag{3.10}$$

   Inserting this into the Fourier integral produces the power spectrum

$$S_{XX}(\omega) = \int_{-\infty}^{\infty} [R_{RR}(\tau) + R_{II}(\tau)][\cos(\omega\tau) - \mathrm{j}\sin(\omega\tau)] + [R_{RI}(\tau) - R_{IR}(\tau)][\mathrm{j}\cos(\omega\tau) + \sin(\omega\tau)]\,\mathrm{d}\tau \tag{3.11}$$

   The product of the sum of the two autocorrelation functions and the sine gives an odd result and consequently does not contribute to the integral. Using Equation (2.48), the difference $R_{RI}(\tau) - R_{IR}(\tau)$ can be rewritten as $R_{RI}(\tau) - R_{RI}(-\tau)$. This is an odd function and multiplied by a cosine the result remains odd. Thus, this product does not contribute to the integral either. Since all imaginary parts cancel out on integration, the resulting power spectrum will be real.

4. Property 4 follows immediately from the definition of Equation (3.2) and the definition of the autocorrelation function (see Equation (2.18)).

Example 3.1:

Consider once more the stochastic process $X(t) = A\cos(\omega_0 t - \Theta)$, with A and ω_0 constants and Θ a random variable that is uniformly distributed on the interval $(0, 2\pi]$. We know that this process is often met in practice. The autocorrelation function of this process has been

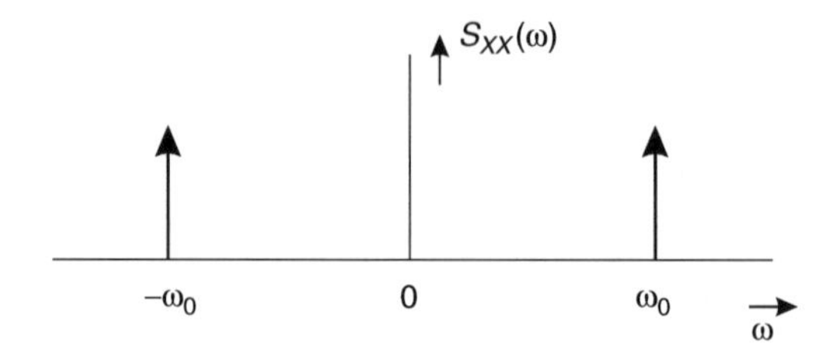

Figure 3.2 The power spectrum of a random phased cosine

shown to be $R_{XX}(\tau) = \frac{1}{2}A^2\cos(\omega_0\tau)$ (see Example 2.1). From a table of Fourier transforms (see Appendix G) it is easily revealed that

$$S_{XX}(\omega) = \frac{\pi}{2}A^2[\delta(\omega - \omega_0) + \delta(\omega + \omega_0)] \tag{3.12}$$

This spectrum has been depicted in Figure 3.2 and consists of two δ functions, one at $\omega = \omega_0$ and another one at $\omega = -\omega_0$. Since the phase is random, introducing an extra constant phase to the cosine does not have any effect on the result. Thus, instead of the cosine a sine wave could also have been taken.

□

Example 3.2:

The second example is also important from a practical point of view, namely the spectrum of an oscillator. From physical considerations the process can be written as $X(t) = A\cos[\omega_0 t + \Psi(t)]$, with A and ω_0 constants and $\Psi(t)$ a random walk process defined by $\Psi(t) = \int_{-\infty}^{t} N(\tau)\mathrm{d}\tau$, where $N(t)$ is a so-called white noise process; i.e. the spectrum of $N(t)$ has a constant value for all frequencies. It can be shown that the autocorrelation function of the process $X(t)$ is [8]

$$R_{XX}(\tau) = \frac{A^2}{2}\exp(-\alpha|\tau|)\cos(\omega_0\tau) \tag{3.13}$$

where ω_0 is the nominal angular frequency of the oscillator and the exponential is due to random phase fluctuations. This autocorrelation function is shown in Figure 3.3(a). It will be clear that A is determined by the total power of the oscillator and from the Fourier table (see Appendix G) the power spectrum

$$S_{XX}(\omega) = \frac{\alpha A^2/2}{\alpha^2 + (\omega - \omega_0)^2} + \frac{\alpha A^2/2}{\alpha^2 + (\omega + \omega_0)^2} \tag{3.14}$$

follows. This spectrum has been depicted in Figure 3.3(b) and is called a Lorentz profile.

□

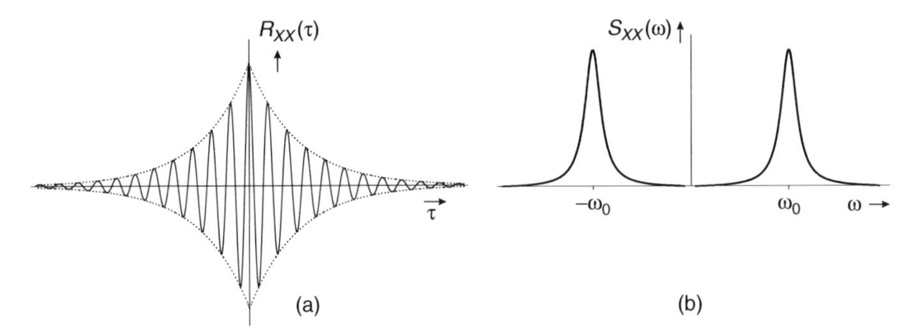

Figure 3.3 (a) The autocorrelation function and (b) the power spectrum of an oscillator

3.2 THE BANDWIDTH OF A STOCHASTIC PROCESS

The r.m.s. bandwidth W_e of a stochastic process is defined using the second normalized moment of the power spectrum, i.e.

$$W_e^2 \triangleq \frac{\int_{-\infty}^{\infty} \omega^2 S_{XX}(\omega)\, d\omega}{\int_{-\infty}^{\infty} S_{XX}(\omega)\, d\omega} \tag{3.15}$$

This definition is, in its present form, only used for lowpass processes, i.e. processes where $S_{XX}(\omega)$ has a significant value at $\omega = 0$ and at low frequencies, and decreasing values of $S_{XX}(\omega)$ at increasing frequency.

Example 3.3:

In this example we will calculate the r.m.s. bandwidth of a very simple power spectrum, namely an ideal lowpass spectrum defined by

$$S_{XX}(\omega) = \begin{cases} 1, & \text{for } |\omega| < B \\ 0, & \text{for } |\omega| \geq B \end{cases} \tag{3.16}$$

Inserting this into the definition of Equation (3.15) yields

$$W_e^2 = \frac{\int_{-B}^{B} \omega^2\, d\omega}{\int_{-B}^{B} d\omega} = \frac{1}{3} B^2 \tag{3.17}$$

The r.m.s. bandwidth is in this case $W_e = B/\sqrt{3}$. This bandwidth might have been expected to be equal to B; the difference is explained by the quadratic weight in the numerator with respect to frequency.

□

In case of bandpass processes (see Subsection 4.4.1) the second, central, normalized moment is used in the definition

$$W_{\mathrm{e}}^{2} \triangleq \frac{4\int_{0}^{\infty}(\omega-\overline{\omega}_{0})^{2}S_{XX}(\omega)\,\mathrm{d}\omega}{\int_{0}^{\infty}S_{XX}(\omega)\,\mathrm{d}\omega} \tag{3.18}$$

where the mean frequency $\overline{\omega}_0$ is defined by

$$\overline{\omega}_{0} \triangleq \frac{\int_{0}^{\infty}\omega S_{XX}(\omega)\,\mathrm{d}\omega}{\int_{0}^{\infty}S_{XX}(\omega)\,\mathrm{d}\omega} \tag{3.19}$$

the first normalized moment of $S_{XX}(\omega)$. A bandpass process is a process where the power spectral density function is confined around a frequency $\bar{\omega}_0$ and which has a negligible value (zero or almost zero) at $\omega = 0$.

The necessity of the factor of 4 in Equation (3.18) compared to Equation (3.15) is explained by the next example.

Example 3.4:

In this example we will consider the r.m.s. bandwidth of an ideal bandpass process with the power spectrum

$$S_{XX}(\omega) = \begin{cases} 1, & \text{for } |\omega-\omega_0| < \dfrac{B}{2} \text{ and } |\omega+\omega_0| < \dfrac{B}{2} \\ 0, & \text{elsewhere} \end{cases} \tag{3.20}$$

The r.m.s. bandwidth follows from the definition of Equation (3.18):

$$W_{\mathrm{e}}^{2} = \frac{4\int_{\omega_0-B/2}^{\omega_0+B/2}(\omega-\omega_0)^{2}\,\mathrm{d}\omega}{\int_{\omega_0-B/2}^{\omega_0+B/2}\mathrm{d}\omega} = \frac{1}{3}B^{2} \tag{3.21}$$

which reveals that the r.m.s. bandwidth equals $W_{\mathrm{e}} = B/\sqrt{3}$. Both the spectrum of the ideal lowpass process from Example 3.3 and the spectrum of the ideal bandpass process from this example are presented in Figure 3.4. From a physical point of view the two processes should

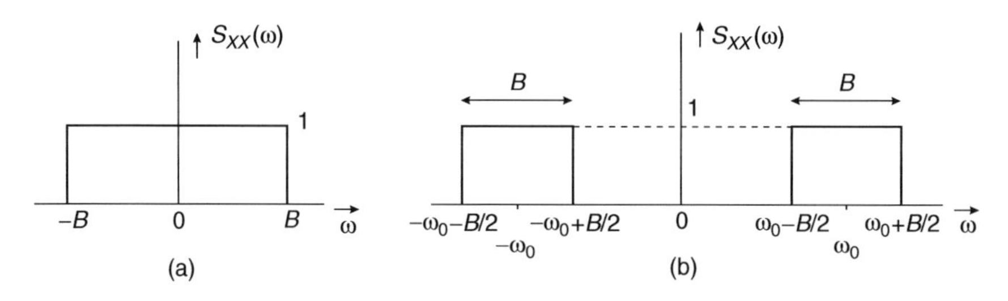

Figure 3.4 (a) Power spectrum of the ideal lowpass process and (b) the power spectrum of the ideal bandpass process

have the same bandwidth and indeed both Equations (3.17) and (3.21) have the same outcome. This is only the case if the factor of 4 is present in Equation (3.18).

□

3.3 THE CROSS-POWER SPECTRUM

Analogous to the preceding section, we can define the cross-power spectral density function, or briefly the cross-power spectrum, as the Fourier transform of the cross-correlation function

$$S_{XY}(\omega) = \int_{-\infty}^{\infty} R_{XY}(\tau) \exp(-j\omega\tau) \, d\tau \tag{3.22}$$

with the corresponding inverse transform

$$R_{XY}(\tau) = \frac{1}{2\pi} \int_{-\infty}^{\infty} S_{XY}(\omega) \exp(j\omega\tau) \, d\omega \tag{3.23}$$

It can be seen that the processes $X(t)$ and $Y(t)$ have to be jointly wide-sense stationary.

A physical interpretation of this spectrum cannot always be given. The function $S_{XY}(\omega)$ often acts as an auxiliary quantity in a few specific problems, such as in bandpass processes (see Section 5.2). Moreover, it plays a role when two (or even more) signals are added. Let us consider the process $Z(t) = X(t) + Y(t)$; then the autocorrelation is

$$R_{ZZ}(\tau) = R_{XX}(\tau) + R_{YY}(\tau) + R_{XY}(\tau) + R_{YX}(\tau) \tag{3.24}$$

From this latter equation the total power of $Z(t)$ is $P_{XX} + P_{YY} + P_{XY} + P_{YX}$ and it follows that the process $Z(t)$ contains, in general, more power than the sum of the powers of the individual signals. This apparently originates from the correlation of the signals. The cross-power spectra show how the additional power components P_{XY} and P_{YX} are spread over the different frequencies, namely

$$P_{XY} \triangleq \frac{1}{2\pi} \int_{-\infty}^{\infty} S_{XY}(\omega) \, d\omega \tag{3.25}$$

and

$$P_{YX} \triangleq \frac{1}{2\pi} \int_{-\infty}^{\infty} S_{YX}(\omega) \, d\omega \tag{3.26}$$

The total amount of additional power may play an important role in situations where an information-carrying signal has to be processed in the midst of additive noise or interference. Moreover, the cross-power spectrum is used to describe bandpass processes (see Chapter 5).

From Equation (3.24) it will be clear that the power of $Z(t)$ equals the sum of the powers in $X(t)$ and $Y(t)$ if the processes $X(t)$ and $Y(t)$ are orthogonal.

Properties of $S_{XY}(\omega)$ for real processes

1. $S_{XY}(\omega) = S_{YX}(-\omega) = S^*_{YX}(\omega)$ (3.27)
2. $\text{Re}\{S_{XY}(\omega)\}$ and $\text{Re}\{S_{YX}(\omega)\}$ are even functions of ω (3.28)

   $\text{Im}\{S_{XY}(\omega)\}$ and $\text{Im}\{S_{YX}(\omega)\}$ are odd functions of ω (3.29)

   where $\text{Re}\{\cdot\}$ and $\text{Im}\{\cdot\}$ are the real and imaginary parts, respectively, of the quantity in the braces
3. If $X(t)$ and $Y(t)$ are independent, then

$$S_{XY}(\omega) = S_{YX}(-\omega) = 2\pi\overline{X}\,\overline{Y}\delta(\omega) \tag{3.30}$$

4. If $X(t)$ and $Y(t)$ are orthogonal, then

$$S_{XY}(\omega) \equiv S_{YX}(\omega) \equiv 0 \tag{3.31}$$

5. If $X(t)$ and $Y(t)$ are uncorrelated, then

$$S_{XY}(\omega) = S_{YX}(-\omega) = 2\pi\overline{X}\,\overline{Y}\delta(\omega) \tag{3.32}$$

Proofs of the properties:

1. Property 1 is proved by invoking Equation (2.48) and the definition of the cross-power spectrum

$$\begin{aligned} S_{XY}(\omega) &= \int_{-\infty}^{\infty} R_{XY}(\tau)\exp(-\mathrm{j}\omega\tau)\,\mathrm{d}\tau = \int_{-\infty}^{\infty} R_{XY}(-\tau)\exp(\mathrm{j}\omega\tau)\,\mathrm{d}\tau \\ &= \int_{-\infty}^{\infty} R_{YX}(\tau)\exp(\mathrm{j}\omega\tau)\,\mathrm{d}\tau = S_{YX}(-\omega) \end{aligned} \tag{3.33}$$

   and from this latter line it follows also that $S_{YX}(-\omega) = S^*_{YX}(\omega)$.

2. In contrast to $S_{XX}(\omega)$ the cross-power spectrum will in general be a complex-valued function. Property 2 follows immediately from the definition

$$S_{XY}(\omega) = \int_{-\infty}^{\infty} R_{XY}(\tau)\cos(\omega\tau)\,\mathrm{d}\tau - \mathrm{j}\int_{-\infty}^{\infty} R_{XY}(\tau)\sin(\omega\tau)\,\mathrm{d}\tau \tag{3.34}$$

   For real processes the cross-correlation function is real as well and the first integral represents the real part of the power spectrum and is obviously even. The second integral represents the imaginary part of the power spectrum which is obviously odd.

3. From Equation (2.56) it is concluded that in this case $R_{XY}(\tau) = \overline{X}\,\overline{Y}$ and its Fourier transform equals the right-hand member of Equation (3.30).

4. The fourth property is quite straightforward. For orthogonal processes, by definition, $R_{XY}(\tau) = R_{YX}(\tau) \equiv 0$, and so are the corresponding Fourier transforms.

5. In the given situation, from Equations (2.69) and (2.72) it is concluded that $R_{XY}(\tau) = \overline{X}\,\overline{Y}$. Fourier transform theory says that the transform of a constant is a δ function of the form given by Equation (3.32).

3.4 MODULATION OF STOCHASTIC PROCESSES

In many applications (such as in telecommunications) a situation is often met where signals are modulated and synchronously demodulated. In those situations the signal is applied to a multiplier circuit, while a second input of the multiplier is a harmonic signal (sine or cosine waveform), called the carrier (see Figure 3.5). We will analyse the spectrum of the output process $Y(t)$ when the spectrum of the input process $X(t)$ is known. In doing so we will assume that the cosine function of the carrier has a constant frequency ω_0, but a random phase Θ that is uniformly distributed on the interval $(0,2\pi]$ and is independent of $X(t)$. The output process is then written as

$$Y(t) = X(t)A_0 \cos(\omega_0 t - \Theta) \tag{3.35}$$

The amplitude A_0 of the carrier is supposed to be constant. The autocorrelation function of the output $Y(t)$ is found by applying the definition to this latter expression, yielding

$$R_{YY}(t, t+\tau) = A_0^2 \mathrm{E}[X(t) \cos(\omega_0 t - \Theta) X(t+\tau) \cos(\omega_0 t + \omega_0 \tau - \Theta)] \tag{3.36}$$

At the start of this chapter we stated that we will confine our analysis to wide-sense stationary processes. We will invoke this restriction for the input process $X(t)$; however, this does not guarantee that the output process $Y(t)$ is also wide-sense stationary. Therefore we used the notation $R_{YY}(t, t+\tau)$ in Equation (3.36) and not $R_{YY}(\tau)$. Elaborating Equation (3.36) yields

$$\begin{aligned} R_{YY}(t, t+\tau) &= \frac{A_0^2}{2} R_{XX}(\tau)\, \mathrm{E}[\cos(2\omega_0 t + \omega_0 \tau - 2\Theta) + \cos \omega_0 \tau] \\ &= \frac{A_0^2}{2} R_{XX}(\tau) \left[\frac{1}{2\pi} \int_0^{2\pi} \cos(2\omega_0 t + \omega_0 \tau - 2\theta)\, \mathrm{d}\theta + \cos \omega_0 \tau \right] \\ &= \frac{A_0^2}{2} R_{XX}(\tau) \cos \omega_0 \tau \end{aligned} \tag{3.37}$$

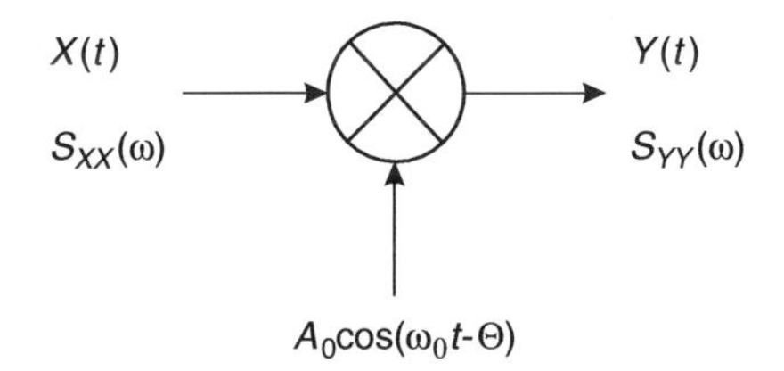

Figure 3.5 A product modulator or mixer

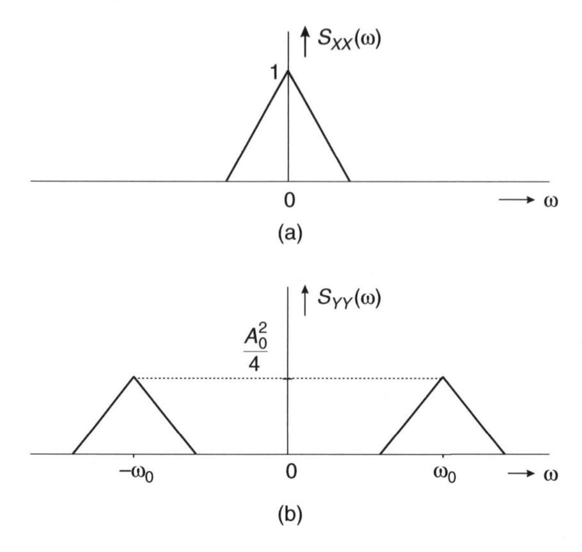

Figure 3.6 The spectra at (a) input ($S_{XX}(\omega)$) and (b) output ($S_{YY}(\omega)$) of a product modulator

From Equation (3.37) it is seen that $R_{YY}(t, t+\tau)$ is independent of t. The mean value of the output is calculated using Equation (3.35); since Θ and $X(t)$ have been assumed to be independent the mean equals the product of the mean values $\mathrm{E}[X(t)]$ and $\mathrm{E}[\cos(\omega_0 t - \Theta)]$. From Example 2.1 we know that this latter mean value is zero. Thus it is concluded that the output process $Y(t)$ is wide-sense stationary, since its autocorrelation function is independent of t and so is its mean. Transforming Equation (3.37) to the frequency domain, we arrive at our final result:

$$S_{YY}(\omega) = \frac{A_0^2}{4}[S_{XX}(\omega - \omega_0) + S_{XX}(\omega + \omega_0)] \tag{3.38}$$

In Figure 3.6 an example has been sketched of a spectrum $S_{XX}(\omega)$. Moreover, the corresponding spectrum $S_{YY}(\omega)$ as it appears at the output of the product modulator is presented. In this figure it has been assumed that $X(t)$ is a lowpass process. The analysis can, however, be applied in a similar way to processes with a different character, e.g. bandpass processes. As a consequence of the modulation we observe a shift of the baseband spectrum to the carrier frequency ω_0 and a shift to $-\omega_0$; actually besides a shift there is also a split-up. This is analogous to the modulation of deterministic signals, a difference being that when dealing with deterministic signals we use the signal spectrum, whereas when dealing with stochastic processes we have to use the power spectrum.

Example 3.5:

The method of modulation is in practice used for demodulation as well; demodulation in this way is called synchronous or coherent demodulation. The basic idea is that multiplication of

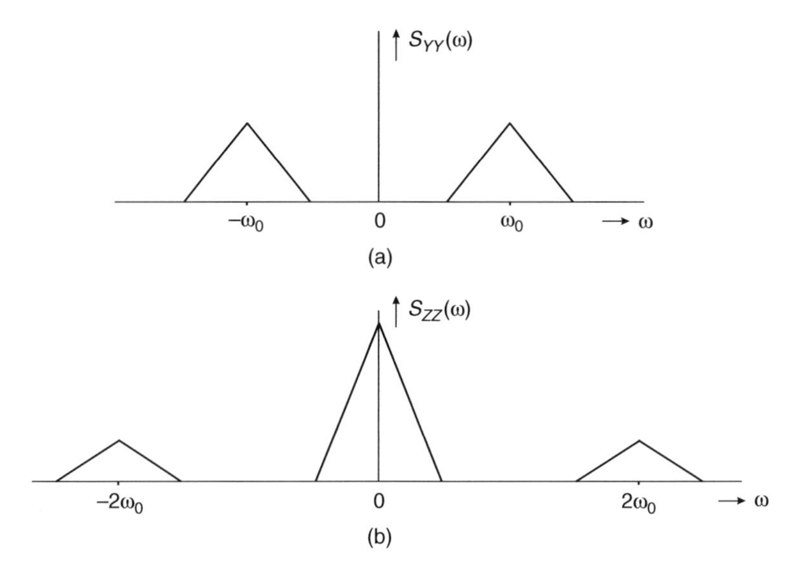

Figure 3.7 (a) The spectra of a modulated signal and (b) the output of the corresponding signal after synchronous demodulation by a product modulator

a signal with an harmonic wave (sine or cosine) shifts the power spectrum by an amount equal to the frequency of the harmonic signal. This shift is twofold: once to the right and once to the left over the frequency axis. Let us apply this procedure to the spectrum of the modulated signal as given in Figure 3.6(b). This figure has been redrawn in Figure 3.7(a). When this spectrum is both shifted to the right and to the left and added, the result is given by Figure 3.7(b). The power spectrum of the demodulated signal consists of three parts:

1. A copy of the original spectrum about $-2\omega_0$;
2. A copy of the original spectrum about $2\omega_0$;
3. Two copies about $\omega = 0$.

The first two copies may be removed by a lowpass filter, whereas the copies around zero actually represent the recovered baseband signal from Figure 3.6(a).

□

Besides modulation and demodulation, multiplication may also be applied for frequency conversion. Modulation and demodulation are therefore examples of frequency conversion (or frequency translation) that can be achieved by using multipliers.

3.4.1 Modulation by a Random Carrier

In certain systems stochastic processes are used as the carrier for modulation. An example is pseudo noise sequences that are used in CDMA (Code Division Multiple Access) systems [9]. The spectrum of such pseudo noise sequences is much wider than that of the modulating signal. A second example is a lightwave communication system, where sources like light emitting diodes (LEDs) also have a bandwidth much wider than the modulating signal. These wideband sources can be described as stochastic processes and we shall denote them by $Z(t)$. For the modulation we use the scheme of Figure 3.5, where the sinusoidal carrier is replaced by this $Z(t)$. If the modulating process is given by $X(t)$, then the modulation signal at the output reads

$$Y(t) = X(t)\,Z(t) \tag{3.39}$$

Assuming that both processes are wide-sense stationary, the autocorrelation function of the output is written as

$$\begin{aligned} R_{YY}(t, t+\tau) &= \mathrm{E}[X(t)Z(t)\,X(t+\tau)Z(t+\tau)] \\ &= \mathrm{E}[X(t)X(t+\tau)\,Z(t)Z(t+\tau)] \end{aligned} \tag{3.40}$$

It is reasonable to assume that the processes $X(t)$ and $Z(t)$ are independent. Then it follows that

$$R_{YY}(t, t+\tau) = R_{XX}(\tau)\,R_{ZZ}(\tau) \tag{3.41}$$

which means that the output is wide-sense stationary as well. Transforming Equation (3.41) to the frequency domain produces the power spectrum of the modulation

$$S_{YY}(\omega) = \frac{1}{2\pi} S_{XX}(\omega) * S_{ZZ}(\omega) \tag{3.42}$$

where $*$ presents the convolution operation.

When $Z(t)$ has a bandpass characteristic and $X(t)$ is a baseband signal, then the modulated signal will be shifted to the bandpass frequency range of the noise-like carrier signal. It is well known that convolution exactly adds the spectral extent of the spectra of the individual signals when they are strictly band-limited. In the case of signals with unlimited spectral extent, the above relation holds approximately in terms of bandwidths [7]. This means that, for example, in the case of CDMA the spectrum of the transmitted signal is much wider than that of the information signal. Therefore, this modulation is also called the spread spectrum technique. On reception, a synchronized version of the pseudo noise signal is generated and synchronous demodulation recovers the information signal. De-spreading is therefore performed in the receiver.

3.5 SAMPLING AND ANALOGUE-TO-DIGITAL CONVERSION

In modern systems extensive use is made of digital signal processors (DSPs), due to the fact that these processors can be programmed and in this way can have a flexible functionality. Moreover, the speed is increasing to such high values that the devices become suitable for many practical applications. However, most signals to be processed are still analogue, such as signals from sensors and communication systems. Therefore sampling of the analogue signal is needed prior to analogue-to-digital (A/D) conversion. In this section we will consider both the sampling process and A/D conversion.

3.5.1 Sampling Theorems

First we will recall the well-known sampling theorem for deterministic signals [7,10], since we need it to describe a sampling theorem for stochastic processes.

> ***Theorem 4***
>
> Suppose that the deterministic signal $f(t)$ has a band-limited Fourier transform $F(\omega)$; i.e. $F(\omega) = 0$ for $|\omega| > W$. Then the signal can exactly be recovered from its samples, if the samples are taken at a sampling rate of at least $1/T_s$, where
>
> $$T_s = \frac{\pi}{W} \tag{3.43}$$
>
> The reconstruction of $f(t)$ from its samples is given by
>
> $$f(t) = \sum_{n=-\infty}^{n=\infty} f(nT_s) \frac{\sin W(t - nT_s)}{W(t - nT_s)} \tag{3.44}$$

The minimum sampling frequency $1/T_s = W/\pi = 2F$ is called the Nyquist frequency, where $F = W/(2\pi)$ is the maximum signal frequency component corresponding to the maximum angular frequency W.

The sampling theorem is understood by considering ideal sampling of the signal. Ideal sampling is mathematically described by multiplying the continuous signal $f(t)$ by an equidistant sequence of δ pulses as was mentioned in Chapter 1. This multiplication is equivalent to a convolution in the frequency domain. From Appendix G it is seen that an infinite sequence of δ pulses in the time domain is in the frequency domain an infinite sequence of δ pulses as well. A sampling rate of $1/T_s$ in the time domain gives a distance of $2\pi/T_s$ between adjacent δ pulses in the frequency domain. This means that in the frequency domain the spectrum $F(\omega)$ of the signal is reproduced infinitely many times shifted over $n2\pi/T_s$, with n an integer running from $-\infty$ to ∞. This is further explained by means of Figure 3.8. From this figure the reconstruction and minimum sampling rate is also understood; namely the original spectrum $F(\omega)$, and thus the signal $f(t)$, is recovered from the

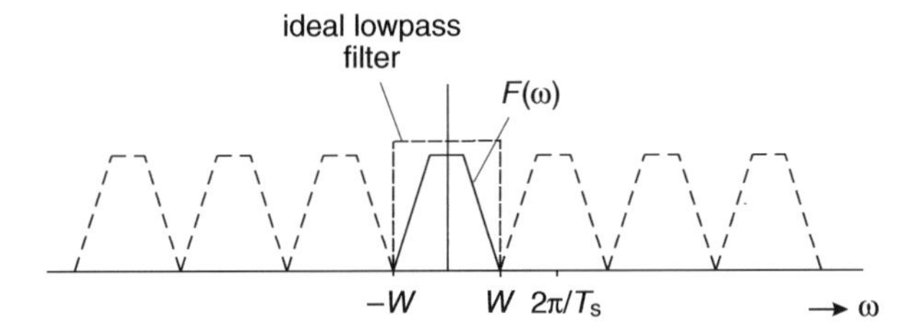

Figure 3.8 The spectrum of a sampled signal and its reconstruction

periodic spectrum by applying ideal lowpass filtering to it. However, ideal lowpass filtering in the time domain is described by a sinc function, as given in Equation (3.44). This sinc function provides the exact interpolation in the time domain. When the sampling rate is increased, the different replicas of $F(\omega)$ become further apart and this will still allow lowpass filtering to filter out $F(\omega)$, as is indicated in Figure 3.9(a). However, decreasing the sampling rate below the Nyquist rate introduces overlap of the replicas (see Figure 3.9(b)) and thus distortion; i.e. the original signal $f(t)$ can no longer be exactly recovered from its samples. This distortion is called aliasing distortion and is depicted in Figure 3.9(c).

For stochastic processes we can formulate a similar theorem.

Theorem 5

Suppose that the wide-sense stationary process $X(t)$ has a band-limited power spectrum $S_{XX}(\omega)$; i.e. $S_{XX}(\omega) = 0$ for $|\omega| > W$. Then the process can be recovered from its samples in the mean-squared error sense, if the samples are taken at a sampling rate of at least $1/T_s$, where

$$T_s = \frac{\pi}{W} \tag{3.45}$$

The reconstruction of $X(t)$ from its samples is given by

$$\hat{X}(t) = \sum_{n=-\infty}^{n=\infty} X(nT_s) \frac{\sin W(t - nT_s)}{W(t - nT_s)} \tag{3.46}$$

The reconstructed process $\hat{X}(t)$ converges to the original process $X(t)$ in the mean-squared error sense, i.e.

$$\mathrm{E}[\{\hat{X}(t) - X(t)\}^2] = 0 \tag{3.47}$$

Proof:

We start the proof by remarking that the autocorrelation function $R_{XX}(t)$ is a deterministic function with a band-limited Fourier transform $S_{XX}(\omega)$, according to the conditions mentioned in Theorem 5. As a consequence, Theorem 4 may be applied to it. The reconstruction of $R_{XX}(t)$ from its samples is written as

$$R_{XX}(t) = \sum_{n=-\infty}^{n=\infty} R_{XX}(nT_s) \frac{\sin W(t - nT_s)}{W(t - nT_s)} = \sum_{n=-\infty}^{n=\infty} R_{XX}(nT_s)\,\mathrm{sinc}[W(t - nT_s)] \tag{3.48}$$

For the proof we need two expressions that are derived from Equation (3.48). The first one is

$$R_{XX}(t) = \sum_{n=-\infty}^{n=\infty} R_{XX}(nT_s - T_1)\,\mathrm{sinc}[W(t - nT_s + T_1)] \tag{3.49}$$

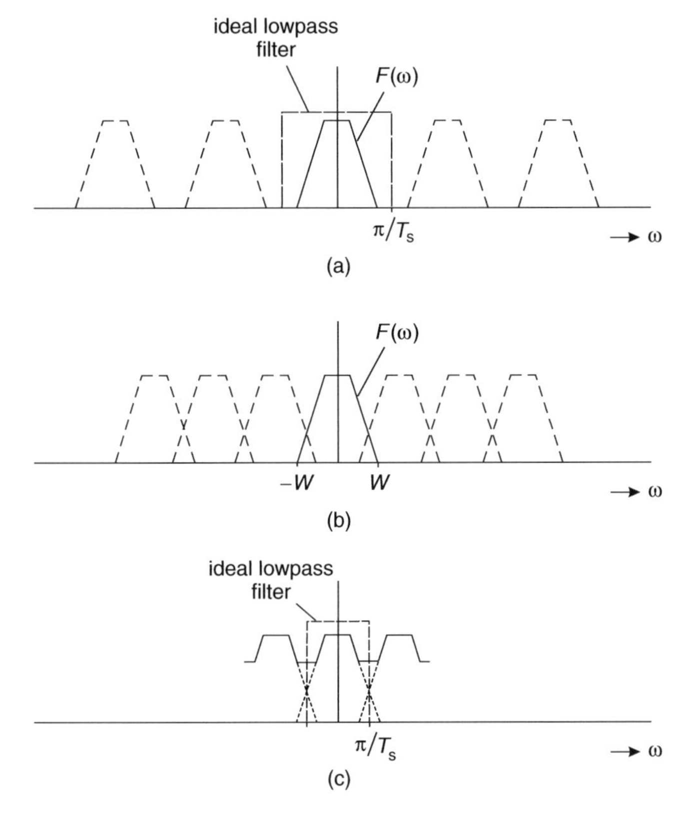

Figure 3.9 The sampled signal spectrum (a) when the sampling rate is higher than the Nyquist rate; (b) when it is lower than the Nyquist rate; (c) aliasing distortion

This expression follows from the fact that the sampling theorem only prescribes a minimum sampling rate, not the exact positions of the samples. The reconstruction is independent of shifting all samples over a certain amount. Another relation we need is

$$\begin{aligned} R_{XX}(t - T_2) &= \sum_{n=-\infty}^{n=\infty} R_{XX}(nT_s)\,\mathrm{sinc}[W(t - T_2 - nT_s)] \\ &= \sum_{n=-\infty}^{n=\infty} R_{XX}(nT_s - T_2)\,\mathrm{sinc}[W(t - nT_s)] \end{aligned} \tag{3.50}$$

The second line above follows by applying Equation (3.49) to the first line.

The mean-squared error between the original process and its reconstruction is written as

$$\begin{aligned} \mathrm{E}[\{\hat{X}(t) - X(t)\}^2] &= \mathrm{E}[\hat{X}^2(t) + X^2(t) - 2\hat{X}(t)X(t)] \\ &= \mathrm{E}[\hat{X}^2(t)] + R_{XX}(0) - 2\mathrm{E}[\hat{X}(t)X(t)] \end{aligned} \tag{3.51}$$

The first term of this latter expression is elaborated as follows:

$$\begin{aligned}
\mathrm{E}[\hat{X}^2(t)] &= \mathrm{E}\left[\sum_{n=-\infty}^{n=\infty} X(nT_s)\,\mathrm{sinc}[W(t-nT_s)] \sum_{m=-\infty}^{m=\infty} X(mT_s)\,\mathrm{sinc}[W(t-mT_s)]\right] \\
&= \sum_{n=-\infty}^{n=\infty} \sum_{m=-\infty}^{m=\infty} \mathrm{E}[X(nT_s)X(mT_s)]\,\mathrm{sinc}[W(t-nT_s)]\,\mathrm{sinc}[W(t-mT_s)] \\
&= \sum_{n=-\infty}^{n=\infty} \left\{ \sum_{m=-\infty}^{m=\infty} R_{XX}(mT_s-nT_s)\,\mathrm{sinc}[W(t-mT_s)] \right\} \mathrm{sinc}[W(t-nT_s)]
\end{aligned} \tag{3.52}$$

To the expression in braces we apply Equation (3.50) with $T_2 = nT_s$. This yields

$$\mathrm{E}[\hat{X}^2(t)] = \sum_{n=-\infty}^{n=\infty} R_{XX}(t-nT_s)\,\mathrm{sinc}[W(t-nT_s)] = R_{XX}(0) \tag{3.53}$$

This equality is achieved when we once more invoke Equation (3.50), but now with $T_2 = t$.
In the last term of Equation (3.51) we insert Equation (3.46). This yields

$$\begin{aligned}
\mathrm{E}[X(t)\hat{X}(t)] &= \mathrm{E}\left[X(t) \sum_{n=-\infty}^{n=\infty} X(nT_s)\,\mathrm{sinc}[W(t-nT_s)]\right] \\
&= \sum_{n=-\infty}^{n=\infty} \mathrm{E}[X(t)\,X(nT_s)]\,\mathrm{sinc}[W(t-nT_s)] \\
&= \sum_{n=-\infty}^{n=\infty} R_{XX}(t-nT_s)\,\mathrm{sinc}[W(t-nT_s)] = R_{XX}(0)
\end{aligned} \tag{3.54}$$

This result follows from Equation (3.53). Inserting Equations (3.53) and (3.54) into Equation (3.51) yields

$$\mathrm{E}[\{\hat{X}(t) - X(t)\}^2] = 0 \tag{3.55}$$

This completes the proof of the theorem.

By means of applying the sampling theorem, continuous stochastic processes can be converted to discrete-time processes, without any information getting lost. This facilitates the processing of continuous processes by DSPs.

3.5.2 A/D Conversion

For processing in a computer or DSP the discrete-time process has to be converted to a discrete random sequence. That conversion is called analogue-to-digital (A/D) conversion. In this conversion the continuous sample values have to be converted to a finite set of discrete values; this is called quantization. It is important to realize that this is a crucial step, since this final set is an approximation of the analogue (continuous) samples. As in all

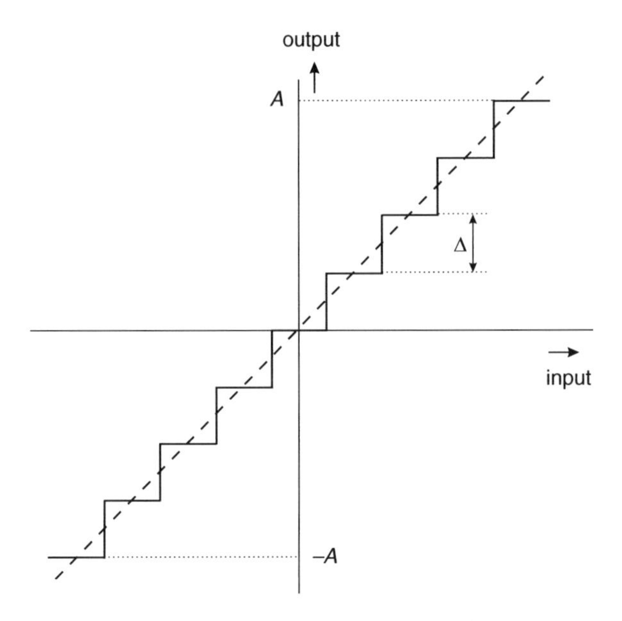

Figure 3.10 Quantization characteristic

approximations there are differences, called errors, between the original signal and the converted one. These errors cannot be restored in the digital-to-analogue reconversion. Thus, it is important to carefully consider the errors.

For the sake of better understanding we will assume that the sample values do not exceed both certain positive and negative values, let us say $|x| \leq A$. Furthermore, we set the number of possible quantization levels equal to $L + 1$. The conversion is performed by a quantizer that has a transfer function as given in Figure 3.10. This quantization characteristic is presented by the solid staircase shaped line, whereas on the dashed line the output is equal to the input. According to this characteristic the quantizer rounds off the input to the closest of the output levels. We assume that the quantizer accommodates the dynamic range of the signal, which covers the range of $\{-A, A\}$. The difference between the two lines represents the quantization error e_q. The mean squared value of this error is calculated as follows. The difference between two successive output levels is denoted by Δ. The exact distribution of the signal between A and $-A$ is, as a rule, not known, but let us make the reasonable assumption that the analogue input values are uniformly distributed between two adjacent levels for all stages. Then the value of the probability density function of the error is $1/\Delta$ and runs from $-\Delta/2$ to $\Delta/2$. The mean value of the error is then zero and the variance

$$\sigma_e^2 = \frac{1}{\Delta}\int_{-\Delta/2}^{\Delta/2} e_\mathrm{q}^2 \mathrm{d}e_\mathrm{q} = \frac{\Delta^2}{12} \tag{3.56}$$

It is concluded that the power of the error is proportional to the square of the quantization step size Δ. This means that this error can be reduced to an acceptable value by selecting an appropriate number of quantization levels. This error introduces noise in the quantization process. By experience it has been established that the power spectral density of the quantization noise extends over a larger bandwidth than the signal bandwidth [11]. Therefore, it behaves approximately as white noise (see Chapter 6).

Example 3.6:

As an example of quantization we consider a sinusoidal signal of amplitude A, half the value of the range of the quantizer. In that case this range is fully exploited and no overload will occur. Recall that the number of output levels was $L+1$, so that the step size is

$$\Delta = \frac{2A}{L} \tag{3.57}$$

Consequently, the power of the quantization error is

$$P_e = \sigma_e^2 = \frac{A^2}{3L^2} \tag{3.58}$$

Remembering that the power in a sinusoidal wave with amplitude A amounts to $A^2/2$, the ratio of the signal power to the quantization noise power follows. This ratio is called the signal-to-quantization noise ratio and is

$$\left(\frac{S}{N}\right)_{\mathrm{q}} = \frac{3L^2}{2} \tag{3.59}$$

When expressing this quantity in decibels (see Appendix B) it becomes

$$10\log\left(\frac{S}{N}\right)_{\mathrm{q}} = 1.8 + 20\log L \quad \mathrm{dB} \tag{3.60}$$

Using binary words of length n to present the different quantization levels, the number of these levels is 2^n. Inserting this into Equation (3.60) it is found that

$$10\log\left(\frac{S}{N}\right)_{\mathrm{q}} \approx 1.8 + 6n \quad \mathrm{dB} \tag{3.61}$$

for large values of L. Therefore, when adding one bit to the word length, the signal-to-quantization noise ratio increases by an amount of 6 dB. For example, the audio CD system uses 16 bits, which yields a signal-to-quantization noise ratio of 98 dB, quite an impressive value.

□

The quantizer characterized by Figure 3.10 is a so-called uniform quantizer, i.e. all steps have the same size. It will be clear that the signal-to-quantization noise ratio can be substantially smaller than the value given by Equation (3.60) if the input amplitude is much smaller than A. This is understood when realizing that in that case the signal power goes down but the quantization noise remains the same.

Non-uniform quantizers have a small step size for small input signal values and this step size increases with increasing input level. As a result the signal-to-quantization noise ratio improves for smaller signal values at the cost of that for larger signal levels (see reference [6]).

After applying sampling and quantization to a continuous stochastic process we actually have a discrete random sequence, but, as mentioned in Chapter 1, these processes are simply special cases of the discrete-time processes.

3.6 SPECTRUM OF DISCRETE-TIME PROCESSES

As usual, the calculation of the power spectrum has to start by considering the autocorrelation function. For the wide-sense stationary discrete-time process $X[n]$ we can write

$$R_{XX}[m] = \mathrm{E}[X[n]\,X[n+m]] \tag{3.62}$$

The relation to the continuous stochastic process is

$$R_{XX}[m] = \sum_{m=-\infty}^{m=\infty} R_{XX}(mT_s)\,\delta(t - mT_s) \tag{3.63}$$

From this equation the power spectral density of $X[m]$ follows by Fourier transformation:

$$S_{XX}(\omega) = \sum_{m=-\infty}^{m=\infty} R_{XX}[m]\,\exp(-\mathrm{j}\omega m T_s) \tag{3.64}$$

Thus, the spectrum is a periodic function (see Figure 3.8) with period $2\pi/T_s$. Such a periodic function can be described by means of its Fourier series coefficients [7,10]

$$R_{XX}[m] = \frac{T_s}{2\pi}\int_{-\pi/T_s}^{\pi/T_s} S_{XX}(\omega)\exp(\mathrm{j}\omega m T_s)\,\mathrm{d}\omega \tag{3.65}$$

In particular, we find for the power of the process

$$P_X = \mathrm{E}[X^2[n]] = R_{XX}[0] = \frac{T_s}{2\pi}\int_{-\pi/T_s}^{\pi/T_s} S_{XX}(\omega)\,\mathrm{d}\omega \tag{3.66}$$

For ease of calculation it is convenient to introduce the z-transform of $R_{XX}[m]$, which is defined as

$$\tilde{S}_{XX}(z) \triangleq \sum_{m=-\infty}^{m=\infty} R_{XX}[m]\,z^{-m} \tag{3.67}$$

The z-transform is more extensively dealt with in Subsection 4.6.2. Comparing this latter expression to Equation (3.64) reveals that the delay operator z equals $\exp(j\omega T_s)$ and consequently

$$\tilde{S}_{XX}(\exp(j\omega m T_s)) = S_{XX}(\omega) \tag{3.68}$$

Example 3.7:

Let us consider a wide-sense stationary discrete-time process $X[n]$ with the autocorrelation sequence

$$R_{XX}[m] = a^{|m|}, \quad \text{with } |a| < 1 \tag{3.69}$$

The spectrum expressed in z-transform notation is

$$\begin{aligned} \tilde{S}_{XX}(z) &= \sum_{m=-\infty}^{m=\infty} a^{|m|} z^{-m} = \sum_{m=-\infty}^{m=-1} a^{-m} z^{-m} + \sum_{m=0}^{m=\infty} a^m z^{-m} \\ &= \frac{az}{1-az} + \frac{z}{z-a} = \frac{1/a - a}{(1/a + a) - (z + 1/z)} \end{aligned} \tag{3.70}$$

From this expression the spectrum in the frequency domain is easily derived by replacing z with $\exp(j\omega T_s)$:

$$S_{XX}(\omega) = \frac{1/a - a}{(1/a + a) - 2\cos(\omega T_s)} \tag{3.71}$$

□

It can be seen that the procedure given here to develop the autocorrelation sequence and spectrum of a discrete-time process can equally be applied to derive cross-correlation sequences and corresponding cross-power spectra. We leave further elaboration on this subject to the reader.

3.7 SUMMARY

The power spectral density function, or power spectrum, of a stochastic process is defined as the Fourier transform of the autocorrelation function. This spectrum shows how the total power of the process is distributed over the various frequencies. Definitions have been given of the bandwidth of stochastic processes. It appears that on modulation the power spectrum is split up into two parts of identical shape as the original unmodulated spectrum: one part is concentrated around the modulation frequency and the other part around minus the modulation frequency. This implies that we use a description based on double-sided spectra. This is very convenient from a mathematical point of view. From a physical point of view negative frequency values have no meaning. When changing to the physical interpretation, the contributions of the power spectrum at negative frequencies are mirrored with respect to the y axis and the values are added to the values at the corresponding positive frequencies.

The sampling theorem is redefined for stochastic processes. Therefore continuous stochastic processes can be converted into discrete-time processes without information becoming lost. For processing signals using a digital signal processor (DSP), still another step is needed, namely analogue-to-digital conversion. This introduces errors that cannot be restored in the digital-to-analogue reconstruction. These errors are calculated and expressed in terms of the signal-to-noise ratio. Finally, the autocorrelation sequence and power spectrum of discrete-time processes are derived.

3.8 PROBLEMS

3.1 A wide-sense stationary process $X(t)$ has the autocorrelation function

$$R_{XX}(\tau) = A\exp\left(-\frac{|\tau|}{T}\right)$$

(a) Calculate the power spectrum $S_{XX}(\omega)$.

(b) Calculate the power of $X(t)$ using the power spectrum.

(c) Check the answer to (b) using Equation (3.8), i.e. based on the autocorrelation function evaluated at $\tau = 0$.

(d) Use Matlab to plot the spectrum for $A = 3$ and $T = 4$.

3.2 Consider the process $X(t)$, of which the autocorrelation function is given in Problem 2.14.

(a) Calculate the power spectrum $S_{XX}(\omega)$. Make a plot of it using Matlab.

(b) Calculate the power of $X(t)$ using the power spectrum.

(c) Calculate the mean value of $X(t)$ from the power spectrum.

(d) Calculate the variance of $X(t)$ using the power spectrum.

(e) Check the answers to (b), (c) and (d) using the answers to Problem 2.14.

(f) Calculate the lowest frequency where the spectrum becomes zero. By means of Matlab calculate the relative amount of a.c. power in the frequency band between zero and this first null.

3.3 A wide-sense stationary process has a power spectral density function

$$S_{XX}(\omega) = \begin{cases} 2, & 0 \le |\omega| < 10/(2\pi) \\ 0, & \text{elsewhere} \end{cases}$$

(a) Calculate the autocorrelation function $R_{XX}(\tau)$.

(b) Use Matlab to plot the autocorrelation function.

(c) Calculate the power of the process, both via the power spectrum and the autocorrelation function.

3.4 Consider the process $Y(t) = A^2 \sin^2(\omega_0 t - \Theta)$, with A and ω_0 constants and Θ a random variable that is uniformly distributed on $(0, 2\pi]$. In Problem 2.8 we calculated its autocorrelation function.

(a) Calculate the power spectrum $S_{YY}(\omega)$.

(b) Calculate the power of $Y(t)$ using the power spectrum.

(c) Check the answer to (b) using Equation (3.8).

3.5 Reconsider Problem 2.9 and insert $\omega_0 = 2\pi$.

(a) Calculate the magnitude of a few spectral lines of the power spectrum by means of Matlab.

(b) Based on (a), calculate an approximate value of the total power and check this on the basis of the autocorrelation function.

(c) Calculate and check the d.c. power.

3.6 Answer the same questions as in Problem 3.5, but now for the process given in Problem 2.11 when inserting $T = 1$.

3.7 A and B are random variables. These variables are used to create the process $X(t) = A \cos \omega_0 t + B \sin \omega_0 t$, with ω_0 a constant.

(a) Assume that A and B are uncorrelated, have zero means and equal variances. Show that in this case $X(t)$ is wide-sense stationary.

(b) Derive the autocorrelation function of $X(t)$.

(c) Derive the power spectrum of $X(t)$. Make a sketch of it.

3.8 A stochastic process is given by $X(t) = A \cos(\Omega t - \Theta)$, where A is a real constant, Ω a random variable with probability density function $f_\Omega(\omega)$ and Θ a random variable that is uniformly distributed on the interval $(0,2\pi]$, independent of Ω. Show that the power spectrum of $X(t)$ is

$$S_{XX}(\omega) = \frac{\pi A^2}{2} [f_\Omega(\omega) + f_\Omega(-\omega)]$$

3.9 A and B are real constants and $X(t)$ is a wide-sense stationary process. Derive the power spectrum of the process $Y(t) = A + B\,X(t)$.

3.10 Can each of the following functions be the autocorrelation function of a wide-sense stationary process $X(t)$?

(a) $R_{XX}(\tau) = \delta(\tau)$.

(b) $R_{XX}(\tau) = \text{rect}(\tau)$.

(c) $R_{XX}(\tau) = \text{tri}(\tau)$.

For definitions of the functions $\delta(\cdot)$, $\text{rect}(\cdot)$ and $\text{tri}(\cdot)$ see Appendix E.

3.11 Consider the process given in Problem 3.1. Based on process $X(t)$ of that problem another process $Y(t)$ is produced, such that

$$S_{YY}(\omega) = \begin{cases} S_{XX}(\omega), & |\omega| < c(1/T) \\ 0, & \text{elsewhere} \end{cases}$$

where c is a constant.

(a) Calculate the r.m.s. bandwidth of $Y(t)$.

(b) Consider the consequences, both for $S_{YY}(\omega)$ and the r.m.s. bandwidth, when $c \to \infty$.

3.12 For the process $X(t)$ it is found that $R_{XX}(\tau) = A \exp[-\tau^2/(2\sigma^2)]$, with A and σ positive constants.

(a) Derive the expression for the power spectrum of $X(t)$.

(b) Calculate the r.m.s. bandwidth of $X(t)$.

3.13 Derive the cross-power spectrum of the processes given in Problem 2.9.

3.14 A stochastic process is defined by $W(t) = AX(t) + BY(t)$, with A and B real constants and $X(t)$ and $Y(t)$ jointly wide-sense stationary processes.

(a) Calculate the power spectrum of $W(t)$.

(b) Calculate the cross-power spectra $S_{XW}(\omega)$ and $S_{YW}(\omega)$.

(c) Derive $S_{WW}(\omega)$ when $X(t)$ and $Y(t)$ are orthogonal.

(d) Derive $S_{WW}(\omega)$ when $X(t)$ and $Y(t)$ are independent.

(e) Derive $S_{WW}(\omega)$ when $X(t)$ and $Y(t)$ are independent and have mean values of zero.

(f) Derive $S_{WW}(\omega)$ when $X(t)$ and $Y(t)$ are uncorrelated.

3.15 A wide-sense stationary noise process $N(t)$ has a power spectrum as given in Figure 3.11. This process is added to an harmonic random signal $S(t) = 3\cos(8t - \Phi)$ and the sum $S(t) + N(t)$ is applied to one of the inputs of a product modulator. To the second input of this modulator another harmonic process $X(t) = 2\cos(8t - \Theta)$ is applied. The random variables Φ and Θ are independent, but have the same uniform distribution on the interval $[0, 2\pi)$. Moreover, these random variables are independent of the process $N(t)$. The output of the product modulator is connected to an ideal lowpass filter with a cut-off angular frequency $\omega_c = 5$.

(a) Make a sketch of the spectrum of the output of the modulator.

(b) Sketch the spectrum at the output of the lowpass filter.

(c) Calculate the d.c. power at the output of the filter.

(d) The output signal is defined as that portion of the output due to $S(t)$. The output noise is defined as that portion of the output due to $N(t)$. Calculate the output signal power and the output noise power, and the ratio between the two (called the signal-to-noise ratio).

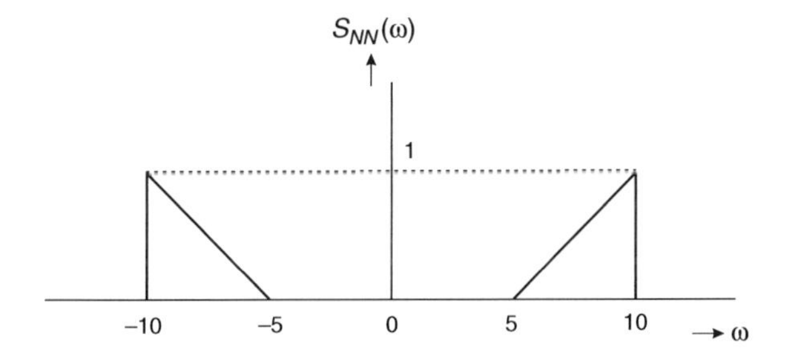

Figure 3.11

3.16 The two independent processes $X(t)$ and $Y(t)$ are applied to a product modulator. Process $X(t)$ is a wide-sense stationary process with power spectral density

$$S_{XX}(\omega) = \begin{cases} 1, & |\omega| \leq W_X \\ 0, & |\omega| > W_X \end{cases}$$

The process $Y(t)$ is an harmonic carrier, but both its phase and frequency are independent random variables

$$Y(t) = \cos(\Omega t - \Theta)$$

where the phase Θ is uniformly distributed on the interval $[0, 2\pi)$ and the carrier frequency is uniformly distributed on the interval $(\omega_0 - W_Y/2 < \Omega < \omega_0 + W_Y/2)$ and with $\omega_0 > W_Y/2 = \text{constant}$.

(a) Calculate the autocorrelation function of the process $Y(t)$.
(b) Is $Y(t)$ wide-sense stationary?
(c) If so, determine and sketch the power spectral density $S_{YY}(\omega)$.
(d) Determine and sketch the power spectral density of the product $Z(t) = X(t)Y(t)$. Assume that $W_Y/2 + W_X < \omega_0$ and $W_X < W_Y/2$.

3.17 The following sequence of numbers represents sample values of a band-limited signal:

$$\ldots, 0, 5, 10, 20, 40, 0, -20, -15, -10, -5, 0, \ldots$$

All other samples are zero.

Use Matlab to reconstruct and graphically represent the signal.

3.18 In the case of ideal sampling, the sampled version of the signal $f(t)$ is represented by

$$f_s(t) = \sum_{n=-\infty}^{n=\infty} f(nT_s)\,\delta(t - nT_s)$$

In so-called 'flat-top sampling' the samples are presented by the magnitude of rectangular pulses, i.e.

$$f_s(t) = \sum_{n=-\infty}^{n=\infty} f(nT_s)\,\text{rect}\left(\frac{t - nT_s}{\tau_s}\right)$$

where $\tau_s < T_s$. (See Appendix E for the definition of the rect$(\cdot)$ function.) Investigate the effect of using these rectangular pulses on the Fourier transform of the recovered signal.

3.19 A voice channel has a spectrum that runs up to 3.4 kHz. On sampling, a guard band (i.e. the distance between adjacent spectral replicas after sampling) of 1.2 kHz has to be taken in account.

(a) What is the minimum sampling rate?

(b) When the samples are coded by means of a linear sampler of 8 bits, calculate the bit rate of a digitized voice channel.

(c) What is the maximum signal-to-noise ratio that can be achieved for such a voice channel?

(d) With how many dB will the signal-to-noise reduce when only half of the dynamic range is used by the signal?

3.20 A stochastic process $X(t)$ has the power spectrum

$$S_{XX} = \frac{1}{1+\omega^2}$$

This process is sampled, and since it is not band-limited the adjacent spectral replicas will overlap. If the spill-over power, i.e. the amount of power that is in overlapping frequency ranges, between adjacent replicas has to be less than 10% of the total power, what is the minimum sampling frequency?

3.21 The discrete-time process $X[n]$ is wide-sense stationary and $R_{XX}[1] = R_{XX}[0]$. Show that $R_{XX}[m] = R_{XX}[0]$ for all m.

3.22 The autocorrelation sequence of a discrete-time wide-sense stationary process $X[n]$ is

$$R_{XX}[m] = \begin{cases} 1 - 0.2|m|, & |m| \leq 4 \\ 0, & |m| > 4 \end{cases}$$

Calculate the spectrum $S_{XX}(\omega)$ and make a plot of it using Matlab.

4

Linear Filtering of Stochastic Processes

In this chapter we will investigate what the influence will be on the main parameters of a stochastic process when filtered by a linear, time-invariant filter. In doing so we will from time to time change from the time domain to the frequency domain and vice versa. This may even happen during the course of a calculation. From Fourier transform theory we know that both descriptions are dual and of equal value, and basically there is no difference, but a certain calculation may appear to be more tractable or simpler in one domain, and less tractable in the other.

In this chapter we will always assume that the input to the linear, time-invariant filter is a wide-sense stationary process, and the properties of these processes will be invoked several times. It should be stressed that the presented calculations and results may only be applied in the situation of wide-sense stationary input processes. Systems that are non-linear or time-variant are not considered, and the same holds for input processes that do not fulfil the requirements for wide-sense stationarity.

We start by summarizing the fundamentals of linear time-invariant filtering.

4.1 BASICS OF LINEAR TIME-INVARIANT FILTERING

In this section we will summarize the theory of continuous linear time-invariant filtering. For the sake of simplicity we consider only single-input single-output (SISO) systems. For a more profound treatment of this theory see references [7] and [10]. The generalization to multiple-input multiple-output (MIMO) systems is straightforward and requires a matrix description.

Let us consider a general system that converts a certain input signal $x(t)$ into the corresponding output signal $y(t)$. We denote this by means of the general hypothetical operator $\mathrm{T}[\cdot]$ as follows (see also Figure 4.1(a)):

$$y(t) = \mathrm{T}[x(t)] \tag{4.1}$$

Introduction to Random Signals and Noise W. van Etten

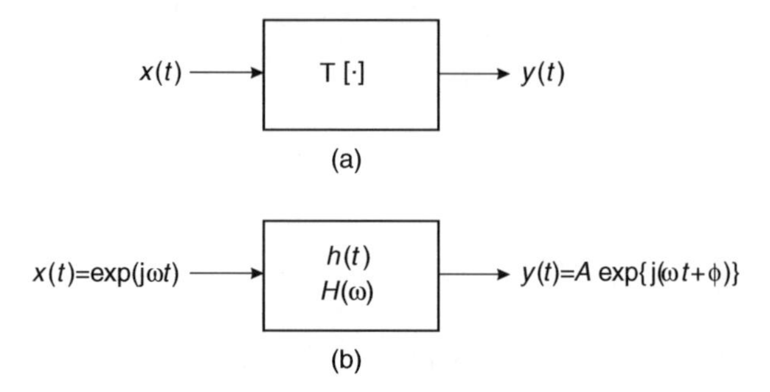

Figure 4.1 (a) General single-input single-output (SISO) system; (b) linear time-invariant (LTI) system

Next we limit our treatment to linear systems and denote this by means of L[·] as follows:

$$y(t) = \mathrm{L}[x(t)] \tag{4.2}$$

The definition of linearity of a system is as follows. Suppose a set of input signals $\{x_n(t)\}$ causes a corresponding set of output signals $\{y_n(t)\}$. Then a system is said to be linear if any arbitrary linear combination of inputs causes the same linear combination of corresponding outputs, i.e.

$$\begin{aligned} &\text{if:} && x_n(t) \Rightarrow y_n(t) \\ &\text{then:} && \sum_n a_n x_n(t) \Rightarrow \sum_n a_n y_n(t) \end{aligned} \tag{4.3}$$

with a_n arbitrary constants. In the notation of Equation (4.2),

$$y(t) = \mathrm{L}\left[\sum_n a_n x_n(t)\right] = \sum_n a_n \mathrm{L}[x_n(t)] = \sum_n a_n y_n(t) \tag{4.4}$$

A system is said to be time-invariant if a shift in time of the input causes a corresponding shift in the output. Therefore,

$$\begin{aligned} &\text{if:} && x_n(t) \Rightarrow y_n(t) \\ &\text{then:} && x_n(t-\tau) \Rightarrow y_n(t-\tau) \end{aligned} \tag{4.5}$$

for arbitrary τ. Finally, a system is linear time-invariant (LTI) if it satisfies both conditions given by Equations (4.3) and (4.5), i.e.

$$\begin{aligned} &\text{if:} && x_n(t) \Rightarrow y_n(t) \\ &\text{then:} && x(t) = \sum_n a_n x_n(t-\tau_n) \Rightarrow y(t) = \sum_n a_n y_n(t-\tau_n) \end{aligned} \tag{4.6}$$

It can be proved [7] that complex exponential time functions, i.e. sine and cosine waves, are so-called eigenfunctions of linear time-invariant systems. An eigenfunction can physically be interpreted as a function that preserves its shape on transmission, i.e. a sine/cosine remains a sine/cosine, but its amplitude and/or phase may change. When these changes are known for all frequencies then the system is completely specified. This specification is done by means of the complex transfer function of the linear time-invariant system. If the system is excited with a complex exponential

$$x(t) = \exp(\mathrm{j}\omega t) \tag{4.7}$$

and the corresponding output is

$$y(t) = A\exp[\mathrm{j}(\omega t + \varphi)] \tag{4.8}$$

with A a real constant, then the transfer function equals

$$H(\omega) = \left.\frac{y(t)}{x(t)}\right|_{x(t)=\exp(\mathrm{j}\omega t)} \tag{4.9}$$

From Equations (4.7) to (4.9) the amplitude and the phase angle of the transfer function follow:

$$\begin{aligned} |H(\omega)| &= A \\ \angle H(\omega) &= \varphi \end{aligned} \tag{4.10}$$

When in Equation (4.9), ω is given all values from $-\infty$ to ∞, the transfer is completely known. As indicated in that equation the amplitude of the input $x(t)$ is taken as unity and the phase zero for all frequencies. All observed complex values of $y(t)$ are then presented by the function $Y(\omega)$. Since the input $X(\omega)$ was taken as unity for all frequencies, Equation (4.9) is in that case written as

$$H(\omega) = \frac{Y(\omega)}{1} \rightarrow Y(\omega) = H(\omega) \times 1 \tag{4.11}$$

From Fourier theory we know that the multiplication in the right-hand side of the latter equation is written in the time domain as a convolution [7,10]. Moreover, the inverse transform of 1 is a δ function. Since a δ function is also called an impulse, the time domain LTI system response following from Equation (4.11) is called the impulse response. We may therefore conclude that the system impulse response $h(t)$ and the system transfer function $H(\omega)$ constitute a Fourier transform pair.

When the transfer function is known, the response of an LTI system to an input signal can be calculated. Provided that the input signal $x(t)$ satisfies the Dirichlet conditions [10], its Fourier transform $X(\omega)$ exists. However, this frequency domain description of the signal is equivalent to decomposing the signal into complex exponentials, which in turn are eigenfunctions of the LTI system. This allows multiplication of $X(\omega)$ by $H(\omega)$ to find $Y(\omega)$, being the Fourier transform of output $y(t)$; namely by taking the inverse transform of $Y(\omega)$, the signal $y(t)$ is reconstructed from its complex exponential components. This justifies the use of Fourier transform theory to be applied to the transmission of signals through LTI signals. This leads us to the following theorem.

Theorem 6

If a linear time-invariant system with an impulse response $h(t)$ is excited by an input signal $x(t)$, then the output is

$$y(t) = \int_{-\infty}^{\infty} h(\tau)\,x(t-\tau)\,\mathrm{d}\tau \tag{4.12}$$

with the equivalent frequency domain description

$$Y(\omega) = H(\omega)\,X(\omega) \tag{4.13}$$

where $X(\omega)$ and $Y(\omega)$ are the Fourier transforms of $x(t)$ and $y(t)$, respectively, and $H(\omega)$ is the Fourier transform of the impulse response $h(t)$.

The two presentations of Equations (4.12) and (4.13) are so-called dual descriptions; i.e. both are complete and either of them is fully determined by the other one. If an important condition for physical realizability of the LTI system is taken into account, namely causality, then the impulse response will be zero for $t < 0$ and the lower bound of the integral in Equation (4.12) changes into zero.

This theorem is the main result we need to describe the filtering of stochastic processes by an LTI system, as is done in the sequel.

4.2 TIME DOMAIN DESCRIPTION OF FILTERING OF STOCHASTIC PROCESSES

Let us now consider the transmission of a stochastic process through an LTI system. Obviously, we may formally apply the time domain description given by Equation (4.12) to calculate the system response of a single realization of the ensemble. However, a frequency domain description is not always possible. Apart from the fact that realizations are often not explicitly known, it may happen that they do not satisfy the Dirichlet conditions. Therefore, we start by characterizing the filtering in the time domain. Later on the frequency domain description will follow from this.

4.2.1 The Mean Value of the Filter Output

The impulse response of the linear, time-invariant filter is denoted by $h(t)$. Let us consider the ensemble of input realizations and call this input process $X(t)$ and the corresponding output process $Y(t)$. Then the relation between input and output is formally described by the convolution

$$Y(t) = \int_{-\infty}^{\infty} h(\rho)X(t-\rho)\,\mathrm{d}\rho \tag{4.14}$$

When the input process $X(t)$ is wide-sense stationary, then the mean value of the output signal is written as

$$\begin{aligned} \mathrm{E}[Y(t)] &= \mathrm{E}\left[\int_{-\infty}^{\infty} h(\rho)X(t-\rho)\,\mathrm{d}\rho\right] = \int_{-\infty}^{\infty} h(\rho)\mathrm{E}[X(t-\rho)]\,\mathrm{d}\rho \\ &= \overline{Y(t)} = \overline{X(t)}\int_{-\infty}^{\infty} h(\rho)\,\mathrm{d}\rho = \overline{X(t)}\,H(0) \end{aligned} \tag{4.15}$$

where $H(\omega)$ is the Fourier transform of $h(t)$. From Equation (4.15) it follows that the mean value of $Y(t)$ equals the mean value of $X(t)$ multiplied by the value of the transfer function for the d.c. component. This value is equal to the area under the curve of the impulse response function $h(t)$. This conclusion is based on the property of $X(t)$ at least being stationary of the first order.

4.2.2 The Autocorrelation Function of the Output

The autocorrelation function of $Y(t)$ is found using the definition of Equation (2.13) and Equation (4.14):

$$\begin{aligned} R_{YY}(t,t+\tau) &= \mathrm{E}[Y(t)\,Y(t+\tau)] \\ &= \mathrm{E}\left[\int_{-\infty}^{\infty} h(\rho_1)X(t-\rho_1)\,\mathrm{d}\rho_1 \int_{-\infty}^{\infty} h(\rho_2)X(t+\tau-\rho_2)\,\mathrm{d}\rho_2\right] \\ &= \iint_{-\infty}^{\infty} \mathrm{E}[X(t-\rho_1)\,X(t+\tau-\rho_2)]h(\rho_1)h(\rho_2)\,\mathrm{d}\rho_1\,\mathrm{d}\rho_2 \end{aligned} \tag{4.16}$$

Invoking $X(t)$ as wide-sense stationary reduces this expression to

$$R_{YY}(\tau) = \iint_{-\infty}^{\infty} R_{XX}(\tau+\rho_1-\rho_2)]h(\rho_1)h(\rho_2)\,\mathrm{d}\rho_1\,\mathrm{d}\rho_2 \tag{4.17}$$

and the mean squared value of $Y(t)$ reads

$$\mathrm{E}[Y^2(t)] = \overline{Y^2(t)} = R_{YY}(0) = \iint_{-\infty}^{\infty} R_{XX}(\rho_1-\rho_2)h(\rho_1)h(\rho_2)\,\mathrm{d}\rho_1\,\mathrm{d}\rho_2 \tag{4.18}$$

From Equations (4.15) and (4.17) it is concluded that $Y(t)$ is wide-sense stationary when $X(t)$ is wide-sense stationary, since neither the right-hand member of Equation (4.15) nor that of Equation (4.17) depends on t.

Equation (4.17) may also be written as

$$R_{YY}(\tau) = R_{XX}(\tau) * h(\tau) * h(-\tau) \tag{4.19}$$

where the symbol $*$ represents the convolution operation.

4.2.3 Cross-Correlation of the Input and Output

The cross-correlation of $X(t)$ and $Y(t)$ is found using Equations (2.46) and (4.14):

$$R_{XY}(t, t+\tau) \triangleq \mathrm{E}[X(t)\,Y(t+\tau)] = \mathrm{E}\left[X(t)\int_{-\infty}^{\infty} h(\rho)X(t+\tau-\rho)\,\mathrm{d}\rho\right]$$
$$= \int_{-\infty}^{\infty} \mathrm{E}[X(t)\,X(t+\tau-\rho)]\,h(\rho)\,\mathrm{d}\rho \tag{4.20}$$

In the case where $X(t)$ is wide-sense stationary Equation (4.20) reduces to

$$R_{XY}(\tau) = \int_{-\infty}^{\infty} R_{XX}(\tau-\rho)h(\rho)\,\mathrm{d}\rho \tag{4.21}$$

This expression may also be presented as the convolution of $R_{XX}(\tau)$ and $h(\tau)$:

$$R_{XY}(\tau) = R_{XX}(\tau) * h(\tau) \tag{4.22}$$

In a similar way the following expression can be derived:

$$R_{YX}(\tau) = \int_{-\infty}^{\infty} R_{XX}(\tau+\rho)h(\rho)\,\mathrm{d}\rho = R_{XX}(\tau) * h(-\tau) \tag{4.23}$$

From Equations (4.21) and (4.23) it is concluded that the cross-correlation functions do not depend on the absolute time parameter t. Earlier we concluded that $Y(t)$ is wide-sense stationary if $X(t)$ is wide-sense stationary. Now we conclude that $X(t)$ and $Y(t)$ are jointly wide-sense stationary if the input process $X(t)$ is wide-sense stationary.

Substituting Equation (4.21) into Equation (4.17) reveals the relation between the autocorrelation function of the output and the cross-correlation between the input and output:

$$R_{YY}(\tau) = \int_{-\infty}^{\infty} R_{XY}(\tau+\rho_1)h(\rho_1)\,\mathrm{d}\rho_1 \tag{4.24}$$

or, presented differently,

$$R_{YY}(\tau) = R_{XY}(\tau) * h(-\tau) \tag{4.25}$$

In a similar way it follows by substitution of Equation (4.23) into Equation (4.17) that

$$R_{YY}(\tau) = \int_{-\infty}^{\infty} R_{YX}(\tau-\rho_2)h(\rho_2)\,\mathrm{d}\rho_2 = R_{YX}(\tau) * h(\tau) \tag{4.26}$$

Example 4.1:

An important application of the cross-correlation function as given by Equation (4.22) consists of the identification of a linear system. If for the input process a white noise process, i.e. a process with a constant value of the power spectral density, let us say of magnitude

$N_0/2$, is selected then the autocorrelation function of that process becomes $N_0\,\delta(\tau)/2$. This makes the convolution very simple, since the convolution of a δ function with another function results in this second function itself. Thus, in that case the cross-correlation function of the input and output yields $R_{XY}(\tau) = N_0\,h(\tau)/2$. Apart from a constant $N_0/2$, the cross-correlation function equals the impulse response of the linear system; in this way we have found a method to measure this impulse response.

□

4.3 SPECTRA OF THE FILTER OUTPUT

In the preceding sections we described the output process of a linear time-invariant filter in terms of the properties of the input process. In doing so we used the time domain description. We concluded that in case of a wide-sense stationary input process the corresponding output process is wide-sense stationary as well, and that the two processes are jointly wide-sense stationary. This offers the opportunity to apply the Fourier transform to the different correlation functions in order to arrive at the spectral description of the output process and the relationship between the input and output processes. It must also be stressed that in this section only wide-sense stationary input processes will be considered.

The first property we are interested in is the spectral density of the output process. Using what has been derived in Section 4.2.2, this is easily revealed by transforming Equation (4.19) to the frequency domain. If we remember that the impulse response $h(\tau)$ is a real function and thus the Fourier transform of $h(-\tau)$ equals $H^*(\omega)$, then the next important statement can be exposed.

Theorem 7

If a wide-sense stationary process $X(t)$, with spectral density $S_{XX}(\omega)$, is applied to the input of a linear, time-invariant filter with the transfer function $H(\omega)$, then the corresponding output process $Y(t)$ is a wide-sense stationary process as well, and the spectral density of the output reads

$$S_{YY}(\omega) = S_{XX}(\omega)\,H(\omega)\,H^*(\omega) = S_{XX}(\omega)\,|H(\omega)|^2 \tag{4.27}$$

The mean power of the output process is written as

$$P_Y = R_{YY}(0) = \frac{1}{2\pi}\int_{-\infty}^{\infty} S_{XX}(\omega)\,|H(\omega)|^2\,\mathrm{d}\omega \tag{4.28}$$

Example 4.2:

Consider the RC network given in Figure 4.2. Then the voltage transfer function of this network is written as

$$H(\omega) = \frac{1}{1 + \mathrm{j}\omega RC} \tag{4.29}$$

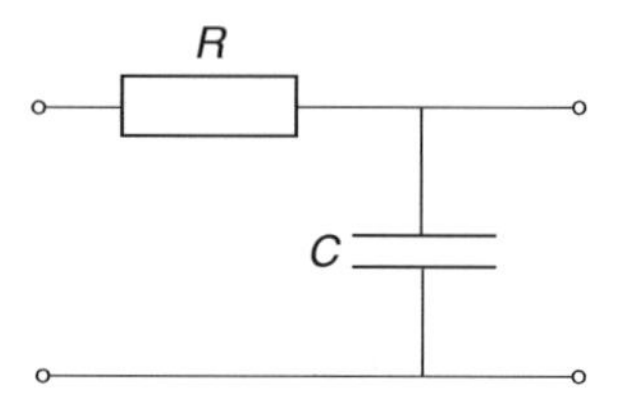

Figure 4.2 RC network

If we assume that the network is excited by a white noise process with spectral density of $N_0/2$ and taking the modulus squared of Equation (4.29), then the output spectral density reads

$$S_{YY}(\omega) = \frac{N_0/2}{1 + (\omega RC)^2} \tag{4.30}$$

For the power in the output process it is found that

$$P_Y = \frac{1}{2\pi} \int_{-\infty}^{\infty} \frac{N_0/2}{1 + (\omega RC)^2} \, d\omega = \frac{N_0}{4\pi RC} \arctan(\omega RC) \Big|_{-\infty}^{\infty} = \frac{N_0}{4RC} \tag{4.31}$$

□

In an earlier stage we found the power of a wide-sense stationary process in an alternative way, namely the value of the autocorrelation function at $\tau = 0$ (see, for instance, Equation (4.16)). The power can also be calculated using that procedure. However, in order to be able to calculate the autocorrelation function of the output $Y(t)$ we need the probability density function of $X(t)$ in order to evaluate a double convolution or we need the probability density function of $Y(t)$. Finding this latter function we meet two main obstacles: firstly, measuring the probability density function is much more difficult than measuring the power density function and, secondly, the probability density function of $Y(t)$ by no means follows in a simple way from that of $X(t)$. This latter statement has one important exception, namely if the probability density function of $X(t)$ is Gaussian then the probability density function of $Y(t)$ is Gaussian as well (see Section 2.3). However, calculating the mean and variance of $Y(t)$, which are sufficient to determine the Gaussian density, using Equations (4.15) and (4.28) is a simpler and more convenient method.

From Equations (4.22) and (4.23) the cross-power spectra are deduced:

$$S_{XY}(\omega) = S_{XX}(\omega)\, H(\omega) \tag{4.32}$$

$$S_{YX}(\omega) = S_{XX}(\omega)\, H(-\omega) = S_{XX}(\omega) H^*(\omega) \tag{4.33}$$

We are now in a position to give the proof of Equation (3.5). Suppose that $S_{XX}(\omega)$ has a negative value for some arbitrary $\omega = \omega_0$. Then a small interval (ω_1, ω_2) about ω_0 is found, such that (see Figure 4.3(a))

$$S_{XX}(\omega) < 0, \quad \text{for } \omega_1 < |\omega| < \omega_2 \tag{4.34}$$

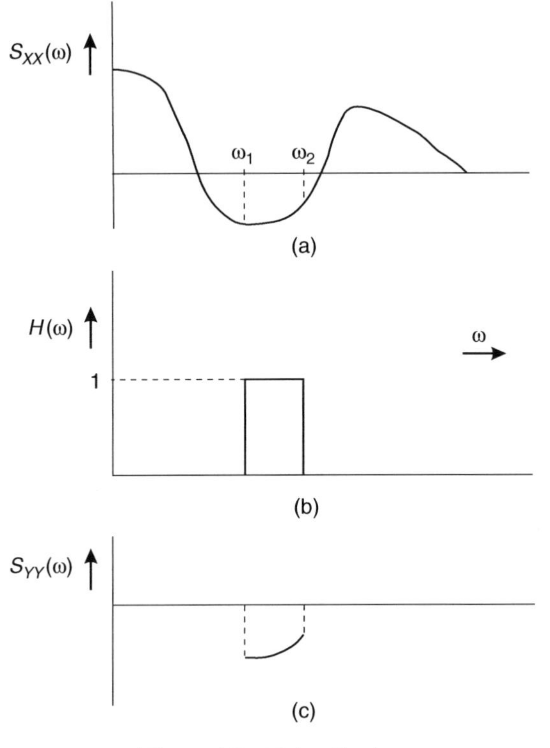

Figure 4.3 Noise filtering

Now consider an ideal bandpass filter with the transfer function (see Figure 4.3(b))

$$H(\omega) = \begin{cases} 1, & \omega_1 < |\omega| < \omega_2 \\ 0, & \text{for all remaining values of } \omega \end{cases} \tag{4.35}$$

If the process $X(t)$, with the power spectrum given in Figure 4.3(a), is applied to the input of this filter, then the spectrum of the output $Y(t)$ is as presented in Figure 4.3(c) and is described by

$$S_{YY}(\omega) = \begin{cases} S_{XX}(\omega), & \omega_1 < |\omega| < \omega_2 \\ 0, & \text{for all remaining values of } \omega \end{cases} \tag{4.36}$$

so that

$$S_{YY}(\omega) \leq 0, \quad \text{for all } \omega \tag{4.37}$$

However, this is impossible as (see Equations (4.28) and (3.8))

$$P_Y = R_{YY}(0) = \frac{1}{2\pi} \int_{-\infty}^{\infty} S_{YY}(\omega)\, d\omega \geq 0 \tag{4.38}$$

This contradiction leads to the conclusion that the starting assumption $S_{XX}(\omega) < 0$ must be wrong.

4.4 NOISE BANDWIDTH

In this section we present a few definitions and concepts related to the bandwidth of a process or a linear, time-invariant system (filter).

4.4.1 Band-Limited Processes and Systems

A process $X(t)$ is called a band-limited process if $S_{XX}(\omega) = 0$ outside certain regions of the ω axis. For a band-limited filter the same definition can be used, provided that $S_{XX}(\omega)$ is replaced by $H(\omega)$. A few special cases of band-limited processes and systems are considered in the sequel.

1. A process is called a lowpass process or baseband process if

$$S(\omega)\begin{cases} \neq 0, & |\omega| < W \\ = 0, & |\omega| > W \end{cases} \tag{4.39}$$

2. A process is called a bandpass process if (see Figure 4.4)

$$S(\omega)\begin{cases} \neq 0, & \omega_0 - W_1 \leq |\omega| \leq \omega_0 - W_1 + W \\ = 0, & \text{for all remaining values of } \omega \end{cases} \tag{4.40}$$

with

$$0 < W_1 < \omega_0 \tag{4.41}$$

3. A system is called a narrowband system if the bandwidth of that system is small compared to the frequency range over which the spectrum of the input process extends. A

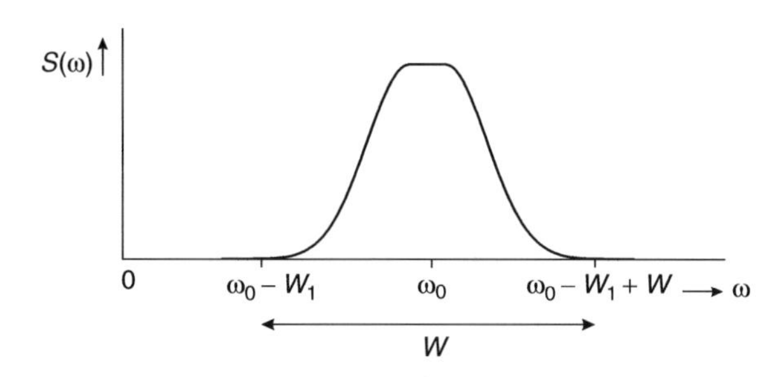

Figure 4.4 The spectrum of a bandpass process (only the region $\omega \geq 0$ is shown)

narrowband bandpass process is a process for which the bandwidth is much smaller than its central frequency, i.e. (see Figure 4.4)

$$W \ll \omega_0 \tag{4.42}$$

The following points should be noted:

- The definitions of processes 1 and 2 can also be used for systems if $S(\omega)$ is replaced by $H(\omega)$.
- In practical systems or processes the requirement that the spectrum or transfer function is zero in a certain region cannot exactly be met in a strict mathematical sense. Nevertheless, we will maintain the given names and concepts for those systems and processes for which the transfer function or spectrum has a negligibly low value in a certain frequency range.
- The spectrum of a bandpass process is not necessarily symmetrical about ω_0.

4.4.2 Equivalent Noise Bandwidth

Equation (4.28) is often used for practical applications. For that reason there is a need for a simplified calculation method in order to compute the noise power at the output of a filter. In this section we will introduce such a simplification.

To that end consider a lowpass system with the transfer function $H(\omega)$. Assume that the spectrum of the input process equals $N_0/2$ for all ω, with N_0 a positive, real constant (such a spectrum is called a white noise spectrum). The power at the output of the filter is calculated using Equation (4.28):

$$P_Y = \frac{1}{2\pi} \int_{-\infty}^{\infty} \frac{N_0}{2} |H(\omega)|^2 \, \mathrm{d}\omega \tag{4.43}$$

Now define an ideal lowpass filter as

$$H_{\mathrm{I}}(\omega) = \begin{cases} H(0), & |\omega| \leq W_{\mathrm{N}} \\ 0, & |\omega| > W_{\mathrm{N}} \end{cases} \tag{4.44}$$

where W_{N} is a positive constant chosen such that the noise power at the output of the ideal filter is equal to the noise power at the output of the original (practical) filter. W_{N} therefore follows from the equation

$$\frac{1}{2\pi} \int_{-\infty}^{\infty} \frac{N_0}{2} |H(\omega)|^2 \, \mathrm{d}\omega = \frac{1}{2\pi} \int_{-W_{\mathrm{N}}}^{W_{\mathrm{N}}} \frac{N_0}{2} |H(0)|^2 \, \mathrm{d}\omega \tag{4.45}$$

If we consider $|H(\omega)|^2$ to be an even function of ω, then solving Equation (4.45) for W_{N} yields

$$W_{\mathrm{N}} = \frac{\int_0^{\infty} |H(\omega)|^2 \mathrm{d}\omega}{|H(0)|^2} \tag{4.46}$$

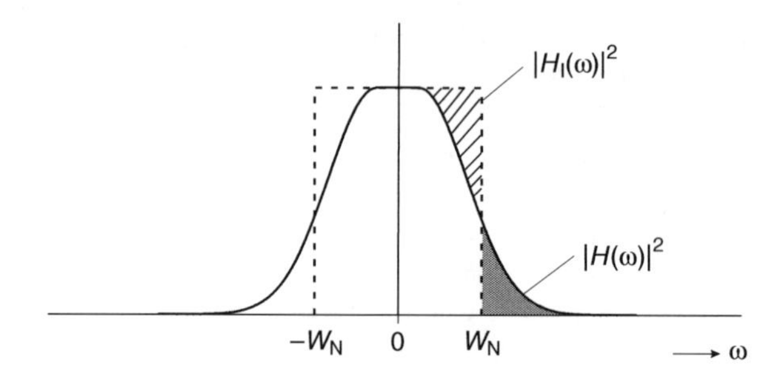

Figure 4.5 Equivalent noise bandwidth of a filter characteristic

W_N is called the equivalent noise bandwidth of the filter with the transfer function $H(\omega)$. In Figure 4.5 it is indicated graphically how the equivalent noise bandwidth is determined. The solid curve represents the practical characteristic and the dashed line the ideal rectangular one. The equivalent noise bandwidth is such that in this picture the dashed area equals the shaded area.

From Equations (4.43) and (4.45) it follows that the output power of the filter can be written as

$$P_Y = \frac{N_0}{2\pi} |H(0)|^2 W_N \tag{4.47}$$

Thus, it can be shown for the special case of white input noise that the integral of Equation (4.43) is reduced to a product and the filter can be characterized by means of a single number W_N as far as the noise filtering behaviour is concerned.

Example 4.3:

As an example of the equivalent noise bandwidth let us again consider the RC network presented in Figure 4.2. Using the definition of Equation (4.46) and the result of Example 4.2 (Equation (4.29)) it is found that $W_N = \pi/(2RC)$. This differs from the 3 dB bandwidth by a factor of $\pi/2$. It may not come as a surprise that the equivalent noise bandwidth of a circuit differs from the 3 dB bandwidth, since the definitions differ. On the other hand, both bandwidths are proportional to $1/(RC)$.

□

Equation (4.47) can often be used for the output of a narrowband lowpass filter. For such a system, which is analogous to Equation (4.43), the output power reads

$$P_Y \approx \frac{1}{2\pi} \int_{-\infty}^{\infty} S_{XX}(0) |H(\omega)|^2 \, d\omega \tag{4.48}$$

and thus

$$P_Y \approx \frac{S_{XX}(0)}{\pi} |H(0)|^2 W_N \tag{4.49}$$

The above calculation may also be applied to bandpass filters. Then it follows that

$$W_{\mathrm{N}} = \frac{\int_0^\infty |H(\omega)|^2 \, \mathrm{d}\omega}{|H(\omega_0)|^2} \tag{4.50}$$

Here ω_0 is a suitably chosen but arbitrary frequency in the passband of the bandpass filter, for instance the centre frequency or the frequency where $|H(\omega)|$ attains its maximum value. The noise power at the output is written as

$$P_Y = \frac{N_0}{2\pi} |H(\omega_0)|^2 W_{\mathrm{N}} \tag{4.51}$$

When once again the input spectrum is approximately constant within the passband of the filter, which often happens in narrowband bandpass filters, then

$$P_Y \approx \frac{S_{XX}(\omega_0)}{\pi} |H(\omega_0)|^2 W_{\mathrm{N}} \tag{4.52}$$

In this way we end up with rather simple expressions for the noise output of linear time-invariant filters.

4.5 SPECTRUM OF A RANDOM DATA SIGNAL

This subject is dealt with here since for the derivation we need results from the filtering of stochastic processes, as dealt with in the preceding sections of this chapter. Let us consider the random data signal

$$X(t) = \sum_n A[n] \, p(t - nT) \tag{4.53}$$

where the data sequence is produced by making a random selection out of the possible values of $A[n]$ for each moment of time nT. In the binary case, for example, we may have $A[n] \epsilon \{-1, +1\}$. The sequence $A[n]$ is supposed to be wide-sense stationary, where $A[n]$ and $A[k]$ in general will be correlated according to

$$\mathrm{E}\big[A[n]A[k]\big] = \mathrm{E}\big[A[n]A[n+m]\big] = \mathrm{E}\big[A[n]A[n-m]\big] = R[m] \tag{4.54}$$

The data symbols amplitude modulate the waveform $p(t)$. This waveform may extend beyond the boundaries of a bit interval. The random data signal $X(t)$ constitutes a cyclo-stationary process. We define a random variable Θ which is uniformly distributed on the interval $(0, T]$ and which is supposed to be independent of the data $A[n]$; this latter assumption sounds reasonable. Using this random variable and the process $X(t)$ we define the new process

$$\underline{X}(t) \triangleq X(t - \Theta) = \sum_n A[n] \, p(t - nT - \Theta) \tag{4.55}$$

Invoking Theorem 1 (see Section 2.2.2) we may conclude that this latter process is stationary. We model the process $\underline{X}(t)$ as resulting from exciting a linear time-invariant system having the impulse response $h(t) = p(t)$ by the input process

$$Y(t) = \sum_n A[n]\, \delta(t - nT - \Theta) \tag{4.56}$$

The autocorrelation function of $Y(t)$ is

$$\begin{aligned} R_{YY}(\tau) &= \mathrm{E}[Y(t)\, Y(t+\tau)] \\ &= \mathrm{E}\left[\sum_n \sum_k A[n]A[k]\, \delta(t - nT - \Theta)\, \delta(t - kT + \tau - \Theta)\right] \\ &= \sum_n \sum_k \mathrm{E}[A[n]\, A[k]]\, \mathrm{E}[\delta(t - nT - \Theta)\, \delta(t - kT + \tau - \Theta)] \\ &= \sum_n \sum_k R[k-n]\, \frac{1}{T} \int_0^T \delta(t - nT - \theta)\, \delta(t + \tau - kT - \theta)\, \mathrm{d}\theta \end{aligned} \tag{4.57}$$

For all values of T, t, τ and n there will be only one single value of k for which both $\delta(t - nT - \theta)$ and $\delta(t + \tau - kT - \theta)$ will be found in the interval $0 < \theta \leq T$. This means that actually the integral in Equation (4.57) is the convolution of two δ functions,which is well defined (see reference [7]). Applying the basic definition of δ functions (see Appendix E) yields

$$R_{YY}(\tau) = \sum_m \frac{R[m]}{T}\, \delta(\tau + mT) \tag{4.58}$$

where $m = k - n$. The autocorrelation function of $Y(t)$ is presented in Figure 4.6. Finally, the autocorrelation function of $\underline{X}(t)$ follows:

$$R_{\underline{XX}}(\tau) = R_{YY}(\tau) * h(\tau) * h(-\tau) = R_{YY}(\tau) * p(\tau) * p(-\tau) \tag{4.59}$$

The spectrum of the random data signal is found by Fourier transforming Equation (4.59):

$$S_{\underline{XX}}(\omega) = S_{YY}(\omega)\, |P(\omega)|^2 \tag{4.60}$$

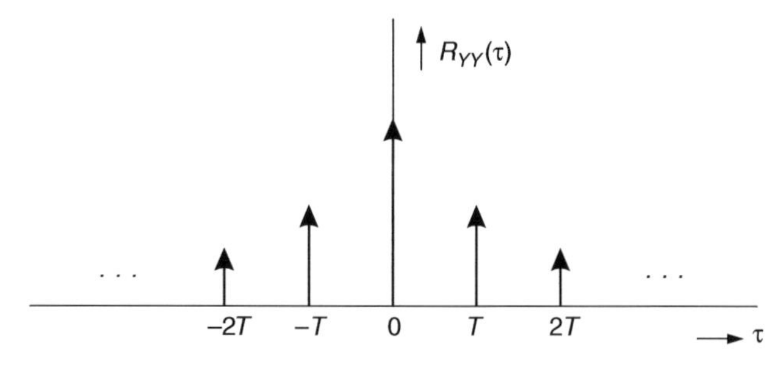

Figure 4.6 The autocorrelation function of the process $Y(t)$ consisting of a sequence of δ functions

where $P(\omega)$ is the Fourier transform of $p(t)$. Using this latter equation and Equation (4.58) the following theorem can be stated.

> ***Theorem 8***
>
> The spectrum of a random data signal reads
>
> $$S_{\underline{XX}}(\omega) = \frac{|P(\omega)|^2}{T} \sum_{m=-\infty}^{\infty} R[m] \exp(\mathrm{j}\omega mT)$$
> $$= \frac{|P(\omega)|^2}{T} \left\{ R[0] + 2\sum_{m=1}^{\infty} R[m] \cos(\omega mT) \right\} \tag{4.61}$$
>
> where $R[m]$ are the autocorrelation values of the data sequence, $P(\omega)$ is the Fourier transform of the data pulses and T is the bit time.

This result, which was found by applying filtering to a stochastic process, is of great importance when calculating the spectrum of digital baseband or digitally modulated signals in communications, as will become clear from the examples in the sequel. When applying this theorem, two cases should clearly be distinguished, since they behave differently, both theoretically and as far as the practical consequences are concerned. We will deal with the two cases by means of examples.

Example 4.4:

The first case we consider is the situation where the mean value of $\underline{X}(t)$ is zero and consequently the autocorrelation function of the sequence $A[n]$ has in practice a finite extent. Let us suppose that the summation in Equation (4.61) runs in that case from 1 to M.

As an important practical example of this case we consider the so-called polar NRZ signal, where NRZ is the abbreviation for non-return to zero, which reflects the behaviour of the data pulses. Possible values of $A[n]$ are $A[n] \in \{+1, -1\}$, where these values are chosen with equal probability and independently of each other. For the signal waveform $p(t)$ we take a rectangular pulse of width T, being the bit time. For the autocorrelation of the data sequence it is found that

$$R[0] = 1^2\tfrac{1}{2} + (-1)^2\tfrac{1}{2} = 1 \tag{4.62}$$

$$R[m] = 1^2\tfrac{1}{4} + (-1)^2\tfrac{1}{4} + (1)(-1)\tfrac{1}{4} + (-1)(1)\tfrac{1}{4} = 0, \quad \text{for } m \neq 0 \tag{4.63}$$

Substituting these values into Equation (4.61) gives the power spectral density of the polar NRZ data signal:

$$S_{\underline{XX}}(\omega) = T\left(\frac{\sin\omega\frac{T}{2}}{\omega\frac{T}{2}}\right)^2 \tag{4.64}$$

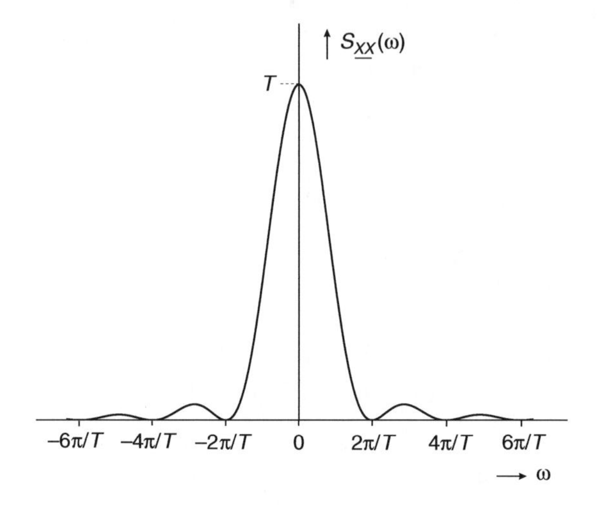

Figure 4.7 The power spectral density of the polar NRZ data signal

since the Fourier transform of the rectangular pulse is the well-known sinc function (see Appendix G). The resulting spectrum is shown in Figure 4.7.

The disadvantage of the polar NRZ signal is that it has a large value of its power spectrum near the d.c. component, although it does not comprise a d.c. component. On the other hand, the signal is easy to generate, and since it is a simplex signal (see Appendix A) it is power efficient.

□

Example 4.5:

In this example we consider the so-called unipolar RZ (return to zero) data signal. This once more reflects the behaviour of the data pulses. For this signal format the values of $A[n]$ are chosen from the set $A[n] \in \{1, 0\}$ with equal probability and mutually independent. The signalling waveform $p(t)$ is defined by

$$p(t) \triangleq \begin{cases} 1, & 0 \leq t < T/2 \\ 0, & T/2 \leq t < T \end{cases} \tag{4.65}$$

It is easy to verify that the autocorrelation of the data sequence reads

$$R[m] = \begin{cases} \frac{1}{2}, & m = 0 \\ \frac{1}{4}, & m \neq 0 \end{cases} \tag{4.66}$$

Inserting this result into Equation (4.61) reveals that we end up with an infinite sum of complex exponentials:

$$S_{\underline{XX}}(\omega) = \frac{T}{4} \left(\frac{\sin \omega \frac{T}{4}}{\omega \frac{T}{4}} \right)^2 \left[\frac{1}{4} + \frac{1}{4} \sum_{m=-\infty}^{\infty} \exp(\mathrm{j}\omega m T) \right] \tag{4.67}$$

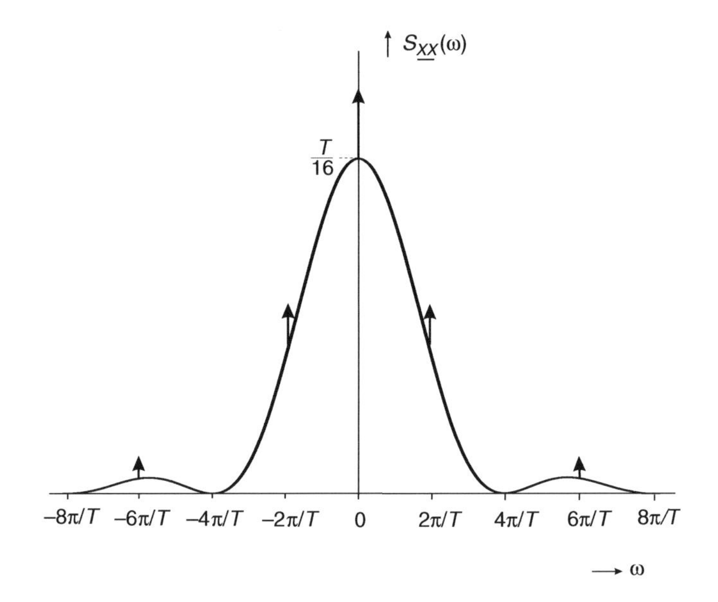

Figure 4.8 The power spectral density of the unipolar RZ data signal

However, this infinite sum of exponentials can be rewritten as

$$\sum_{m=-\infty}^{\infty} \exp(\mathrm{j}\omega mT) = \frac{2\pi}{T} \sum_{m=-\infty}^{\infty} \delta\left(\omega - \frac{2\pi m}{T}\right) \tag{4.68}$$

which is known as the Poisson sum formula [7]. Applying this sum formula to Equation (4.67) yields

$$S_{\underline{XX}}(\omega) = \frac{T}{16}\left(\frac{\sin \omega \frac{T}{4}}{\omega \frac{T}{4}}\right)^2 \left[1 + \frac{2\pi}{T} \sum_{m=-\infty}^{\infty} \delta\left(\omega - \frac{2\pi m}{T}\right)\right] \tag{4.69}$$

This spectrum has been depicted in Figure 4.8. Comparing this figure with that of Figure 4.7 a few remarks need to be made. First of all, the first null bandwidth of the RZ signal increased by a factor of two compared to the NRZ signal. This is due to the fact that the pulse width was reduced by the same factor. Secondly, a series of δ functions appears in the spectrum. This is due to the fact that the unipolar signal has no zero mean. Besides the large value of the spectrum near zero frequency there is a d.c. component. This is also discovered from the δ function at zero frequency. The weights of the δ functions scale with the sinc function (Equation (4.69)) and vanish at all zero-crossings of the latter.

□

Theorem 8 is a powerful tool for calculating the spectrum of all types of formats for data signals. For more spectra of data signals see reference [6].

4.6 PRINCIPLES OF DISCRETE-TIME SIGNALS AND SYSTEMS

The description of discrete-time signals and systems follows in a straightforward manner by sampling the functions that describe their continuous counterparts. Especially for the theory and conversion it is most convenient to confine the procedure to ideal sampling, i.e. by multiplying the continuous functions by a sequence of δ functions, where these δ functions are equally spaced in time.

First of all we define the discrete-time δ function as follows:

$$\delta[n] \triangleq \begin{cases} 1, & n = 0, \\ 0, & n \neq 0 \end{cases} \tag{4.70}$$

or in general

$$\delta[n-m] \triangleq \begin{cases} 1, & n = m \\ 0, & n \neq m \end{cases} \tag{4.71}$$

Based on this latter definition each arbitrary signal $x[n]$ can alternatively be denoted as

$$x[n] = \sum_{m=-\infty}^{\infty} x[m]\,\delta[n-m] \tag{4.72}$$

We will limit our treatment to linear time-invariant systems. For discrete-time systems we introduce a similar definition for linear time-invariant systems as we did for continuous systems, namely

$$\begin{aligned} &\text{if:} && x_i[n] \Rightarrow y_i[n] \\ &\text{then:} && \sum_i a_i x_i[n-m_i] \Rightarrow \sum_i a_i y_i[n-m_i] \end{aligned} \tag{4.73}$$

for any arbitrary set of constants a_i and m_i.

Also, the convolution follows immediately from the continuous case

$$y[n] = \sum_{m=-\infty}^{\infty} x[m]\,h[n-m] = \sum_{m=-\infty}^{\infty} h[n]\,x[n-m] = x[n] * h[n] \tag{4.74}$$

This latter description will be used for the output $y[n]$ of a discrete-time system with the impulse response $h[n]$, which has $x[n]$ as an input signal. This expression is directly deduced by sampling Equation (4.12), but can alternatively be derived from Equation (4.72) and the properties of linear time-invariant systems.

In many practical situations the impulse response function $h[n]$ will have a finite extent. Such filters are called finite impulse response filters, abbreviated as FIR filters, whereas the infinite impulse response filter is abbreviated to the IIR filter.

4.6.1 The Discrete Fourier Transform

For continuous signals and systems we have a dual frequency domain description. Now we look for discrete counterparts for both the Fourier transform and its inverse transform, since then these transformations can be performed by a digital processor. When considering for discrete-time signals the presentation by means of a sequence of δ functions

$$x[n] = \sum_{n=-\infty}^{\infty} x(t)\,\delta(t - nT_s) = \sum_{n=-\infty}^{\infty} x(nT_s)\,\delta(t - nT_s) = \sum_{n=-\infty}^{\infty} x[n]\,\delta(t - nT_s) \tag{4.75}$$

where $1/T_s$ is the sampling rate, the corresponding Fourier transform of the sequence $x[n]$ is easily achieved, namely

$$X(\omega) = \sum_{n=-\infty}^{\infty} x[n]\,\exp(-\mathrm{j}\omega nT_s) \tag{4.76}$$

Due to the discrete character of the time function its Fourier transform is a periodic function of frequency with period $2\pi/T_s$. The inverse transform is therefore

$$x[n] = \frac{T_s}{2\pi}\int_{-\pi/T_s}^{\pi/T_s} X(\omega)\exp(\mathrm{j}n\omega T_s)\,\mathrm{d}\omega \tag{4.77}$$

In Equations (4.76) and (4.77) the time domain has been discretized. Therefore, the operations are called the discrete-time Fourier transform (DTFT) and the inverse discrete-time Fourier transform (IDTFT), respectively. However, the frequency domain has not yet been discretized in those equations. Let us therefore now introduce a discrete presentation of ω as well. We define the radial frequency step as $\Delta\omega \triangleq 2\pi/T$, where T still has to be determined. Moreover, the number of samples has to be limited to a finite amount, let us say N. In order to arrive at a self-consistent discrete Fourier transform this number has to be the same for both the time and frequency domains. Inserting this into Equation (4.76) gives

$$X[k] = X\left(k\frac{2\pi}{T}\right) = \sum_{n=0}^{N-1} x[n]\,\exp\left(-\mathrm{j}k\frac{2\pi}{T}nT_s\right) \tag{4.78}$$

If we define the ratio of T and T_s as $N \triangleq T/T_s$, then

$$X[k] = \sum_{n=0}^{N-1} x[n]\,\exp\left(-\mathrm{j}2\pi\frac{kn}{N}\right) \tag{4.79}$$

Inserting the discrete frequency and limited amount of samples as defined above into Equation (4.77) and approximating the integral by a sum yields

$$x[n] = \frac{T_s}{2\pi}\sum_{k=0}^{N-1} X[k]\,\exp\left(\mathrm{j}2\pi\frac{nk}{T}T_s\right)\frac{2\pi}{T} \tag{4.80}$$

or

$$x[n] = \frac{1}{N}\sum_{k=0}^{N-1} X[k] \exp\left(j2\pi\frac{nk}{N}\right) \tag{4.81}$$

The transform given by Equation (4.79) is called the discrete Fourier transform (abbreviated DFT) and Equation (4.81) is called the inverse discrete Fourier transform (IDFT). They are a discrete Fourier pair, i.e. inserting a sequence into Equation (4.79) and in turn inserting the outcome into Equation (4.81) results in the original sequence, and the other way around. In this sense the deduced set of two equations is self-consistent. It follows from the derivations that they are used as discrete approximations of the Fourier transforms. Modern mathematical software packages such as Matlab comprise special routines that perform the DFT and IDFT. The algorithms used in these packages are called the fast Fourier transform (FFT). The FFT algorithm and its inverse (IFFT) are just DFT, respectively IDFT, but are implemented in a very efficient way. The efficiency is maximum when the number of samples N is a power of 2. For more details on DFT and FFT see references [10] and [12].

Example 4.6:

It is well known from Fourier theory that the transform of a rectangular pulse in the time domain corresponds to a sinc function in the frequency domain (see Appendices E and G). Let us check this result by applying the FFT algorithm of Matlab to a rectangular pulse. Before programming the pulse, a few peculiarities of the FFT algorithm should be observed. First of all, it is only based on the running variables n and k, both running from 0 to $N-1$, which means that no negative values along the x axis can be presented. Here it should be emphasized that in Matlab vectors run from 1 to N, which means that when applying this package the x axis is actually shifted over one position. Another important property is that since the algorithm both requires and produces N data values, Equations (4.79) and (4.81) are periodic functions of respectively k and n, and show a periodicity of N. Thanks to this periodicity the negative argument values can be displayed in the second half of the period. This actually means that in Figure 4.9(a) the rectangular time function is centred about zero. Similarly, the frequency function as displayed in Figure 4.9(b) is actually centred about zero as well. In this figure it has been taken that $N = 256$.

In Figure 4.10 the functions are redrawn with the second half shifted over one period to the left. This results in functions centred about zero in both domains and reflects the well-known result from Fourier theory. It will be clear that the actual time scale and frequency scale in this figure are to be determined based on the actual width of the rectangular pulse.

□

From a theoretical point of view it is impossible for both the time function and the corresponding frequency function to be of finite extent. However, one can imagine that if one of them is limited the parameters of the DFT are chosen such that the transform is a good approximation. Care should be taken with this, as shown in Figure 4.11. In this figure the rectangular pulse in the time domain has been narrowed, which results in a broader function

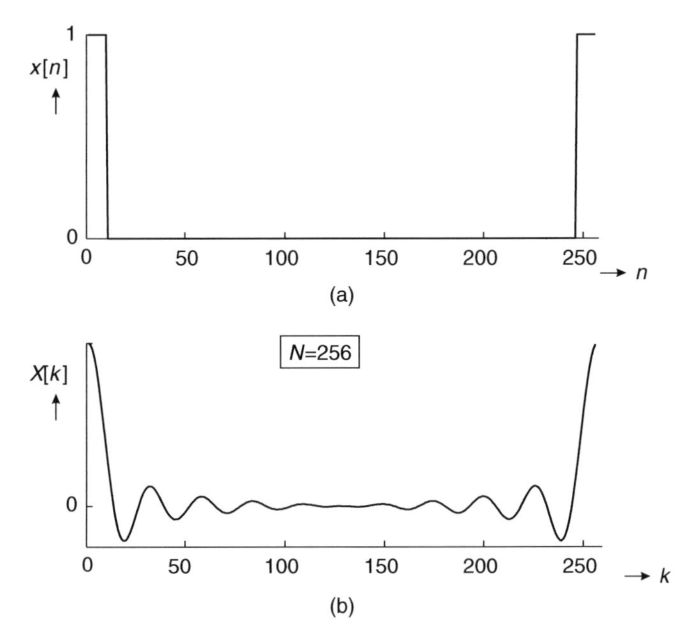

Figure 4.9 (a) The FFT applied to a rectangular pulse in the time domain; (b) the transform in the frequency domain

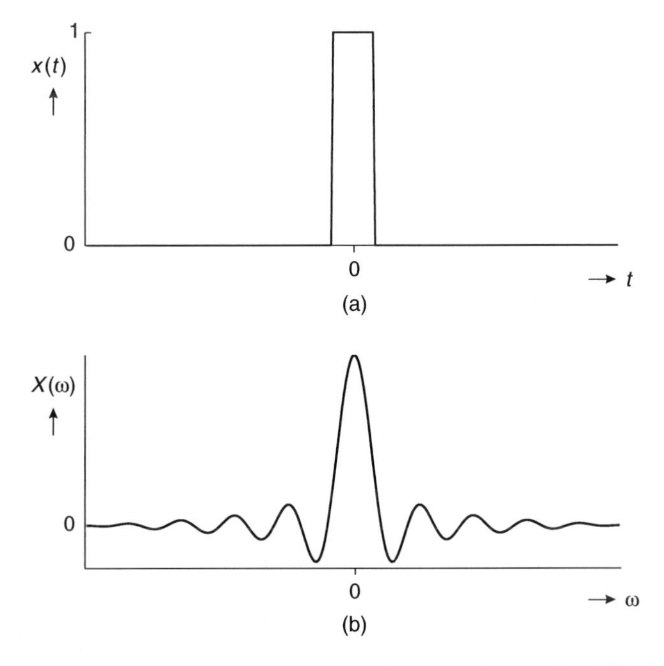

Figure 4.10 (a) The shifted rectangular pulse in the time domain; (b) the shifted transform in the frequency domain

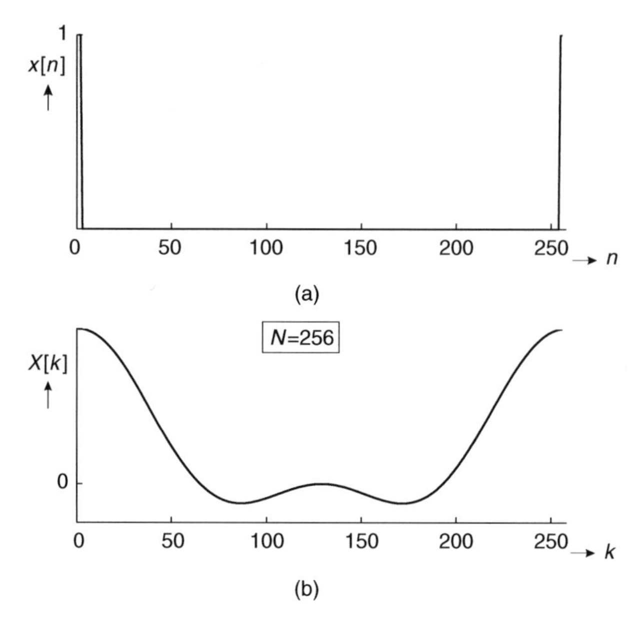

Figure 4.11 (a) Narrowed pulse in the time domain; (b) the FFT result, showing aliasing

in the frequency domain. In this case two adjacent periods in the frequency domain overlap; this is called aliasing.

Although the two functions of Figure 4.11 form an FFT pair, the frequency domain result from Figure 4.11(b) shows a considerable distortion compared to the Fourier transform, which is rather well approximated by Figure 4.9(b). It will be clear that this aliasing can in such cases be prevented by increasing the number of samples N. This has to be done in this case by keeping the number of samples of value 1 the same, and inserting extra zeros in the middle of the interval.

4.6.2 The z-Transform

An alternative approach to deal with discrete-time signals and systems is by setting $\exp(j\omega T_s) \triangleq z$ in Equation (4.76). This results in the so-called z-transform, which is defined as

$$\tilde{X}(z) \triangleq \sum_{n=-\infty}^{\infty} x[n]\, z^{-n} \tag{4.82}$$

Comparing this with Equation (4.76) it is concluded that

$$\tilde{X}(\exp(j\omega T_s)) = X(\omega) \tag{4.83}$$

Since Equation (4.82) is exactly the same as the discrete Fourier transform, only a different notation has been introduced, the same operations used with the Fourier transform are allowed. If we consider Equation (4.82) as the z-transform of the input signal to a linear time-invariant discrete-time system, when calculating the z-transform of the impulse response

$$\tilde{H}(z) = \sum_{n=-\infty}^{\infty} h[n]\, z^{-n} \tag{4.84}$$

the z-transform of the output is found to be

$$\tilde{Y}(z) = \tilde{H}(z)\, \tilde{X}(z) \tag{4.85}$$

A system is called stable if it has a bounded output signal when the input is bounded. A discrete-time system is a stable system if all the poles of $\tilde{H}(z)$ lie inside the unit circle of the z plane or in terms of the impulse response [10]:

$$\sum_{n=-\infty}^{\infty} |h[n]| < \infty \tag{4.86}$$

The z-transform is a very powerful tool to use when dealing with discrete-time signals and systems. This is due to the simple and compact presentation of it on the one hand and the fact that the coefficients of the different powers z^{-n} are identified as the time samples at nT_s on the other hand. For further details on the z-transform see references [10] and [12].

Example 4.7:

Consider a discrete-time system with the impulse response

$$h[n] = \begin{cases} a^n, & n \geq 0 \\ 0, & n < 0 \end{cases} \tag{4.87}$$

and where $|a| < 1$. The sequence $h[n]$ has been depicted in Figure 4.12 for a positive value of a. The z-transform of this impulse response is

$$\tilde{H}(z) = 1 + az^{-1} + a^2 z^{-2} + \cdots = \frac{1}{1 - az^{-1}} \tag{4.88}$$

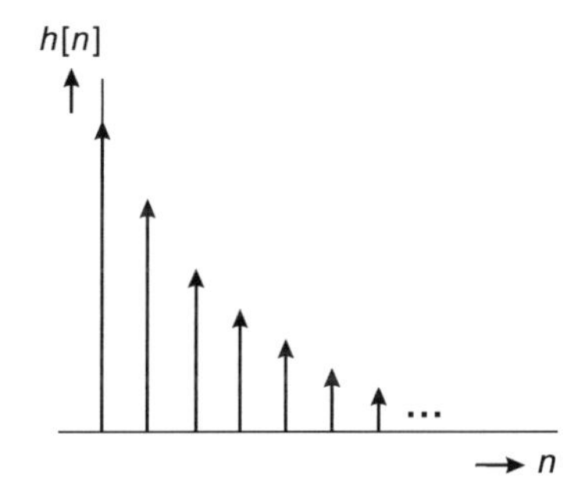

Figure 4.12 The sequence a^n with $0 < a < 1$

Let us suppose that this filter is excited with an input sequence

$$x[n] = \begin{cases} b^n, & n \geq 0 \\ 0, & n < 0 \end{cases} \tag{4.89}$$

with $|b| < 1$ and $b \neq a$. Similar to Equation (4.88) the z-transform of this sequence is

$$\tilde{X}(z) = 1 + bz^{-1} + b^2z^{-2} + \cdots = \frac{1}{1 - bz^{-1}} \tag{4.90}$$

Then the z-transform of the output is

$$\tilde{Y}(z) = \tilde{H}(z)\,\tilde{X}(z) = \frac{1}{1 - az^{-1}}\,\frac{1}{1 - bz^{-1}} \tag{4.91}$$

The time sequence can be recovered from this by decomposition into partial fractions:

$$\frac{1}{1 - az^{-1}}\,\frac{1}{1 - bz^{-1}} = \frac{a}{a - b}\,\frac{1}{1 - az^{-1}} - \frac{b}{a - b}\,\frac{1}{1 - bz^{-1}} \tag{4.92}$$

From this the output sequence is easily derived:

$$y[n] = \begin{cases} \dfrac{a}{a - b}a^n - \dfrac{b}{a - b}b^n, & n \geq 0 \\ 0, & n < 0 \end{cases} \tag{4.93}$$

□

As far as the realization of discrete-time filters is concerned two main types are distinguished, namely the non-recursive filter structure and the recursive. The realization of the non-recursive filter is quite straightforward and the structure is given in Figure 4.13; it is also called the transversal filter or tapped delay line filter. The boxes represent delays of T_s seconds and the outputs are multiplied by a_n. The delayed and multiplied outputs are added

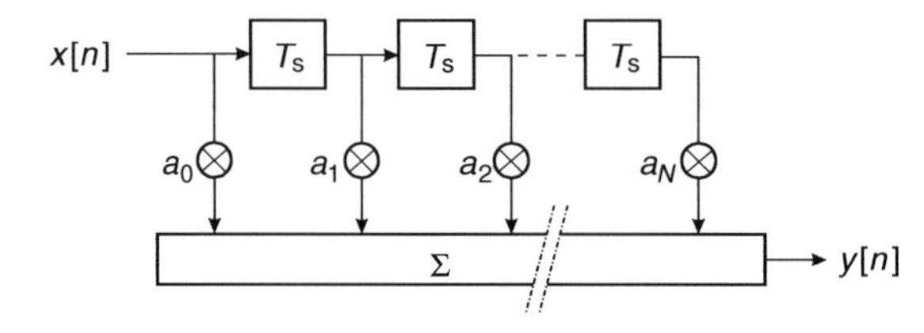

Figure 4.13 The structure of the discrete-time non-recursive filter, transversal filter or tapped delay line filter

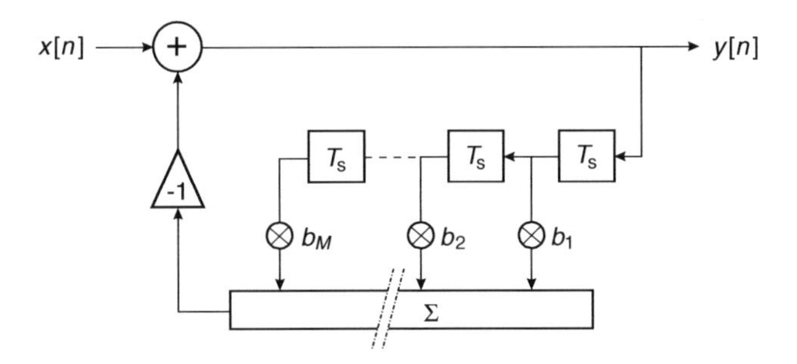

Figure 4.14 The structure of the discrete-time recursive filter

to form the filter output sequence $y[n]$. It is easy to understand that the transfer function of the filter in the z domain is described by the polynomial

$$\tilde{H}(z) = \tilde{A}(z) = a_0 + a_1 z^{-1} + \cdots + a_N z^{-N} = \sum_{n=0}^{n=N} a_n z^{-n} \tag{4.94}$$

From the structure it follows that it is a finite impulse response (FIR) filter and there is a simple and direct relation between the multiplication factors and the polynomial coefficients. An FIR filter is inherently stable.

The recursive filter is based on a similar tapped delay line filter, which is in a feedback loop depicted in Figure 4.14. The transfer function of the feedback filter is

$$\tilde{B}(z) = b_1 z^{-1} + b_2 z^{-2} + \cdots + b_M z^{-M} \tag{4.95}$$

and from the figure it is easily derived that the transfer function of the recursive filter is

$$\tilde{H}(z) \triangleq \frac{\tilde{Y}(z)}{\tilde{X}(z)} = \frac{1}{1 + \tilde{B}(z)} \tag{4.96}$$

As a rule this transfer function has an infinite impulse response, so it is an IIR filter. The stability of the recursive filter is guaranteed if the denominator polynomial $1 + \tilde{B}(z)$ in Equation (4.96) has no zeros outside the unit circle.

In filter synthesis the transfer function is often specified by means of a rational function; i.e. it is given as

$$\tilde{H}(z) = \frac{a_0 + a_1 z^{-1} + a_2 z^{-2} + \cdots + a_N z^{-N}}{1 + b_1 z^{-1} + b_2 z^{-2} + \cdots + b_M z^{-M}} = \frac{\tilde{A}(z)}{1 + \tilde{B}(z)} \tag{4.97}$$

It can be seen that this function is realizable as the cascade of the FIR filter from Figure 4.13 and the IIR filter from Figure 4.14, as follows from Equations (4.94) to (4.96), while the same stability criterion is valid as for the IIR filter.

The Signal Processing Toolbox from Matlab comprises several commands for the analysis and design of discrete-time filters, one of which is `filter`, which calculates the output of filters described by Equation (4.97) when excited by a specified input sequence.

4.7 DISCRETE-TIME FILTERING OF RANDOM SEQUENCES

4.7.1 Time Domain Description of the Filtering

The filtering of a discrete-time stochastic process $X[n]$ by a discrete-time linear time-invariant system with the impulse response $h[n]$ is described by the convolution (see Equation (4.74))

$$Y[n] = \sum_{m=-\infty}^{\infty} X[m]\, h[n-m] = \sum_{m=-\infty}^{\infty} X[n-m]\, h[m] = X[n] * h[n] \tag{4.98}$$

It should be remembered that treatment is confined to real wide-sense stationary processes. Then the mean value of the output sequence is

$$\mathrm{E}\big[Y[n]\big] = \sum_{m=-\infty}^{\infty} \mathrm{E}\big[X[n-m]\big]\, h[m] = \overline{X} \sum_{m=-\infty}^{\infty} h[m] = \overline{X}\tilde{H}(1) \tag{4.99}$$

where $\tilde{H}(1)$ is the z-transform of $h[n]$ evaluated at $z = 1$, which follows from its definition (see Equation (4.84)).

The autocorrelation sequence of the output process, which can be proved to be wide-sense stationary as well, is

$$\begin{aligned}
R_{YY}[m] &= \mathrm{E}[Y[n]\, Y[n+m]] \\
&= \mathrm{E}\left[\sum_{k=-\infty}^{\infty} X[n-k]\, h[k] \sum_{l=-\infty}^{\infty} X[n+m-l]\, h[l]\right] \\
&= \sum_{k=-\infty}^{\infty} \sum_{l=-\infty}^{\infty} \mathrm{E}[X[n-k]\, X[n+m-l]]\, h[k]\, h[l] \\
&= R_{XX}[m] * h[m] * h[-m]
\end{aligned} \tag{4.100}$$

The cross-correlation sequence between the input and output becomes

$$\begin{aligned}
R_{XY}[m] &= \mathrm{E}\big[X[n]\, Y[n+m]\big] \\
&= \mathrm{E}\left[X[n] \sum_{l} X[n+m-l]\, h[l]\right] \\
&= \sum_{l} \mathrm{E}\big[X[n]\, X[n+m-l]\big]\, h[l] \\
&= \sum_{l} R_{XX}[m-l]\, h[l] = R_{XX}[m] * h[m]
\end{aligned} \tag{4.101}$$

In a similar way is derived

$$R_{YX}[m] = R_{XX}[m] * h[-m] \tag{4.102}$$

Moreover, the following relation exists:

$$R_{YY}[m] = R_{XY}[m] * h[-m] = R_{YX}[m] * h[m] \tag{4.103}$$

4.7.2 Frequency Domain Description of the Filtering

Once the autocorrelation sequence at the output of the linear time-invariant discrete-time system is known, the spectrum at the output follows from Equation (4.100):

$$S_{YY}(\omega) = S_{XX}(\omega)\, H(\omega)\, H^*(\omega) = S_{XX}(\omega)\, |H(\omega)|^2 \tag{4.104}$$

where the functions of frequency have to be interpreted as in Equation (4.76).

Using the z-transform, and assuming $h[n]$ to be real, we arrive at

$$\tilde{S}_{YY}(z) = \tilde{S}_{XX}(z)\, \tilde{H}(z)\, \tilde{H}(z^{-1}) \tag{4.105}$$

since for a real system $H^*(\omega) = H(-\omega)$ and consequently $\tilde{H}(z) = \tilde{H}(z^{-1})$.

Owing to the discrete-time nature of $Y[n]$ its spectrum is periodic. According to Equation (3.66) its power is denoted by

$$\begin{aligned} P_Y = \mathrm{E}\left[Y^2[n]\right] = R_{YY}[0] &= \frac{T_s}{2\pi} \int_{-\pi/T_s}^{\pi/T_s} S_{XX}(\omega)\, |H(\omega)|^2\, \mathrm{d}\omega \\ &= \frac{T_s}{2\pi} \int_{-\pi/T_s}^{\pi/T_s} S_{XX}(\omega)\, |\tilde{H}(\exp(\mathrm{j}\omega T_s))|^2\, \mathrm{d}\omega \\ &= \sum_k \sum_l R_{XX}[k-l]\, h[k]\, h[l] \end{aligned} \tag{4.106}$$

The last line of this equation follows from Equation (4.100).

Example 4.8:

Let us consider the system with the z-transform

$$\tilde{H}(z) = \frac{1}{1 - az^{-1}} \tag{4.107}$$

with $|a| < 1$; this is the system introduced in Example 4.7. We assume that the system is driven by the stochastic process $X[n]$ with spectral density $\tilde{S}_{XX}(z) = 1$ or equivalently $R_{XX}[m] = \delta[m]$; later on we will call such a process white noise. Then the output spectral density according to Equation (4.105) is written as

$$\tilde{S}_{YY}(z) = \frac{z}{z-a}\, \frac{z^{-1}}{z^{-1}-a} = \frac{z}{(z-a)(1-az)} \tag{4.108}$$

Expanding this expression in partial fractions yields

$$\tilde{S}_{YY}(z) = \frac{1}{1-a^2}\frac{1}{1-az} + \frac{1}{1-a^2}az^{-1}\frac{1}{1-az^{-1}} \tag{4.109}$$

Next we expand in series the fractions with z and z^{-1}:

$$\begin{aligned}\tilde{S}_{YY}(z) &= \frac{1}{1-a^2}(1 + az + a^2z^2 + \cdots) + \frac{1}{1-a^2}az^{-1}(1 + az^{-1} + a^2z^{-2} + \cdots) \\ &= \frac{1}{1-a^2}\left(\sum_{n=-\infty}^{0} a^{-n}z^{-n} + \sum_{n=1}^{\infty} a^n z^{-n}\right) = \frac{1}{1-a^2}\sum_{n=-\infty}^{\infty} a^{|n|}z^{-n}\end{aligned} \tag{4.110}$$

The autocorrelation sequence of the output is easily derived from this, namely

$$R_{YY}[m] = \frac{1}{1-a^2}a^{|m|} \tag{4.111}$$

and in turn it follows immediately that

$$P_Y = R_{YY}[0] = \frac{1}{1-a^2} \tag{4.112}$$

The power spectrum is deduced from Equation (4.109), which for that purpose is rewritten as

$$\tilde{S}_{YY}(z) = \frac{1}{1 - a(z + z^{-1}) + a^2} \tag{4.113}$$

Inserting $z = \exp(j\omega T_s)$ leads to the spectrum in the frequency domain

$$S_{YY}(\omega) = \tilde{S}_{YY}(\exp(j\omega T_s)) = \frac{1}{1 - 2a\cos(\omega T_s) + a^2} \tag{4.114}$$

In Figure 4.15 this spectrum has been depicted for $a = 0.8$ and a sampling interval time of 1 second. As a consequence the spectrum has a periodicity of 2π in the angular frequency domain. Clearly, this is a lowpass spectrum.

□

From this example it is concluded that the z-transform is a powerful tool for analysing discrete-time systems that are driven by a stochastic process. The transformation and its inverse are quite simple and the operations in the z domain are simply algebraic manipulations. Moreover, conversion from the z domain to the frequency domain is just a simple substitution.

Matlab comprises the procedures `conv` and `deconv` for multiplication and division of polynomials, respectively. The `filter` command in the Signal Processing Toolbox can do the same job (see also the end of Subsection 4.6.2). Moreover, the same toolbox comprises the command `freqz`, which, apart from the filter operation, also converts the result to the frequency domain.

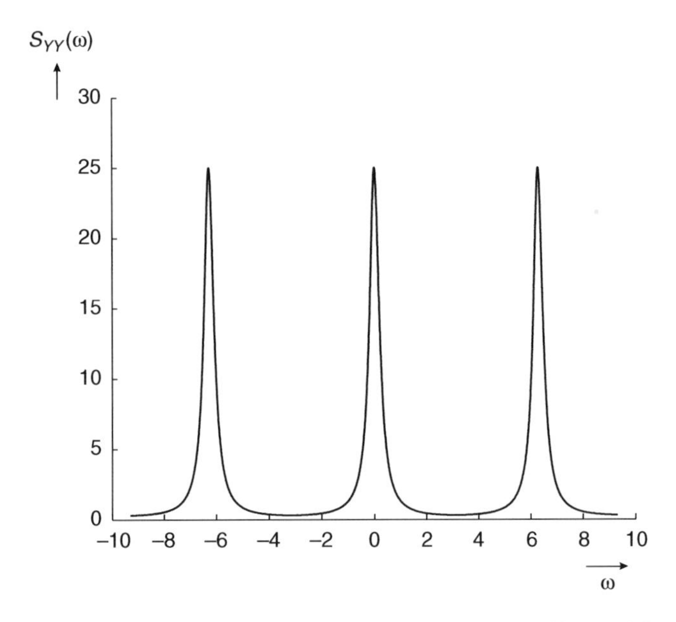

Figure 4.15 The periodic spectrum of the example with $a = 0.8$

4.8 SUMMARY

The input of a linear time-invariant system is excited with a wide-sense stationary process. In the time domain the output process is described as the convolution of the input process and the impulse response of the system, e.g. a filter. Next from this description such characteristics as mean value and correlation functions are to be determined, using the definitions given in Chapter 2. Although the expression for the autocorrelation function of the output process looks rather complicated, namely a twofold convolution, for certain processes this may result in a tractable description. Applying the Fourier transform and arriving at the power spectral density produces a much simpler expression. The probability density function of the output process can, in general, not be easily determined in analytical form from the input probability density function. There is an important exception for this; namely if the input process has a Gaussian probability density function then the output process also has a Gaussian probability density function.

The concept of equivalent noise bandwidth has been defined in order to arrive at an even more simple description of noise filtering in the frequency domain. The theory of noise filtering is applied to a specific stochastic process in order to describe the autocorrelation function and spectrum of random data signals.

Next, attention is paid to discrete-time signals and systems. Specials tools for that are dealt with. The discrete Fourier transform (DFT) and its inverse (IDFT) are derived and it is shown that this transform can serve as an approximation of the Fourier transform. Problems when applying the DFT for that purpose are indicated, including the ways used to avoid them. A closely related transform, namely the z-transform, appears to be more tractable for practical applications. The relation to the Fourier transform is quite simple. Finally, it is shown how to apply these transforms to filtering discrete-time processes by discrete-time systems.

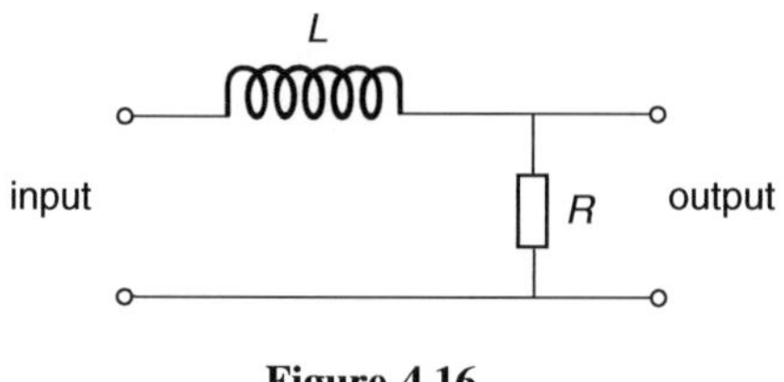

Figure 4.16

4.9 PROBLEMS

4.1 Consider the network given in Figure 4.16.

(a) Calculate the voltage transfer function. Make a plot of its absolute value on a double logarithmic scale using Matlab and with $R/L = 1$.

(b) Calculate the impulse response. Make a plot of it using Matlab.

(c) Calculate the response of the network to a rectangular input pulse that starts at $t = 0$, has height 1 and lasts for 2 seconds. Make a plot of the response using Matlab.

4.2 Derive the Fourier transform of the function

$$f(t) = \exp(-\alpha|t|)\cos(\omega_0 t)$$

Use Matlab to produce a plot of the transform with $\alpha = 1$ and $\omega_0 = 10$.

4.3 Consider the circuit given in Figure 4.17.

(a) Determine the impulse response of the circuit. Make a sketch of it.

(b) Calculate the transfer function $H(\omega)$.

(c) Determine and draw the response $y(t)$ to the input signal $x(t) = A\,\mathrm{rect}[(t - \frac{1}{2}T)/T]$, where A is a constant. (See Appendix E for the definition of the rectangular pulse function rect$(\cdot)$).

(d) Determine and draw the response $y(t)$ to the input signal $x(t) = A\,\mathrm{rect}[(t - T)/(2T)]$, where A is a constant.

4.4 A stochastic process $X(t) = A\sin(\omega_0 t - \Theta)$ is given, with A and ω_0 real, positive constants and Θ a random variable that is uniformly distributed on the interval $(0, 2\pi]$. This process is applied to a linear time-invariant network with the impulse response

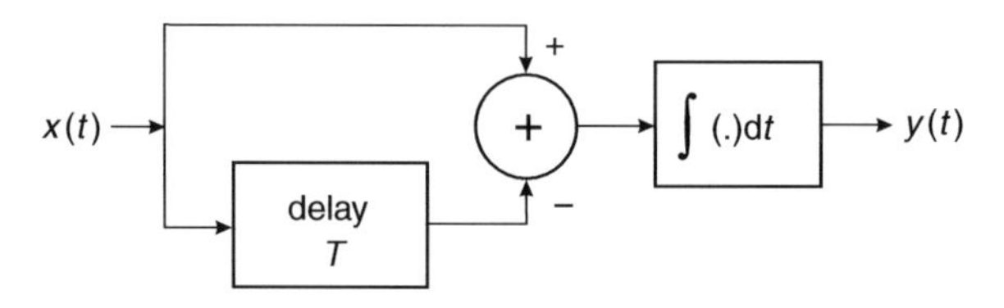

Figure 4.17

$h(t) = \mathrm{u}(t)\exp(-t/\tau_0)$. (The unit-step function $\mathrm{u}(t)$ is defined in Appendix E.) Here τ_0 is a positive, real constant. Derive an expression for the output process.

4.5 A wide-sense stationary Gaussian process with spectral density $N_0/2$ is applied to the input of a linear time-invariant filter. The impulse response of the filter is

$$h(t) = \begin{cases} 1, & 0 < t < T \\ 0, & \text{elsewhere} \end{cases}$$

(a) Sketch the impulse response of the filter.
(b) Calculate the mean value and the variance of the output process.
(c) Determine the probability density function of the output process.
(d) Calculate the autocorrelation function of the output process.
(e) Calculate the power spectrum of the output process.

4.6 A wide-sense process $X(t)$ has the autocorrelation function $R_{XX}(\tau) = A^2 + B(1 - |\tau|/T)$ for $|\tau| < T$ and with A, B and T positive constants. This process is used as the input to a linear, time-invariant system with the impulse response $h(t) = \mathrm{u}(t) - \mathrm{u}(t - T)$.

(a) Sketch $R_{XX}(\tau)$ and $h(t)$.
(b) Calculate the mean value of the output process.

4.7 White noise with spectral density of $N_0/2$ V^2/Hz is applied to the input of the system given in Problem 4.1.

(a) Calculate the spectral density of the output.
(b) Calculate the mean quadratic value of the output process.

4.8 Consider the circuit in Figure 4.18, where $X(t)$ is a wide-sense stationary voltage process. Measurements on the output voltage process $Y(t)$ reveal that this process is Gaussian. Moreover, it is measured as

$$R_{YY}(\tau) = 9\exp(-\alpha|\tau|) + 25, \quad \text{where } \alpha = \frac{1}{RC}$$

(a) Determine the probability density function $f_Y(y)$. Plot this function using Matlab.
(b) Calculate and sketch the spectrum $S_{XX}(\omega)$ of the input process.
(c) Calculate and sketch the autocorrelation function $R_{XX}(\tau)$ of the input process.

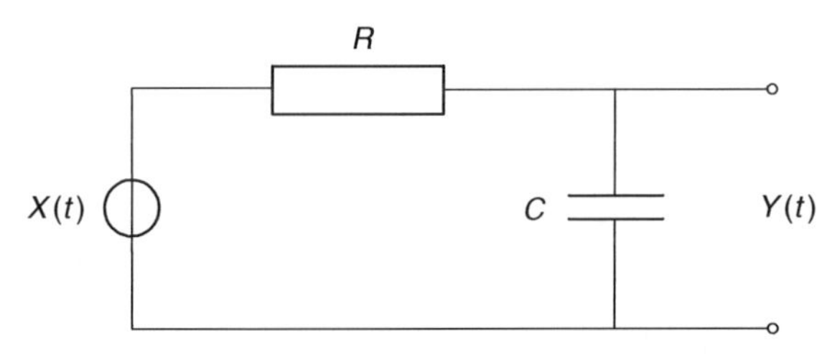

Figure 4.18

4.9 Two linear time-invariant systems have impulse responses of $h_1(t)$ and $h_2(t)$, respectively. The process $X_1(t)$ is applied to the first system and the corresponding response reads $Y_1(t)$. Similarly, the input process $X_2(t)$ to the second system results in the response $Y_2(t)$. Calculate the cross-correlation function of $Y_1(t)$ and $Y_2(t)$ in terms of $h_1(t)$, $h_2(t)$ and the cross-correlation function of $X_1(t)$ and $X_2(t)$, assuming $X_1(t)$ and $X_2(t)$ to be jointly wide-sense stationary.

4.10 Two systems are cascaded. A stochastic process $X(t)$ is applied to the input of the first system having the impulse response $h_1(t)$. The response of this system is $W(t)$ and serves as the input of the second system with the impulse response $h_2(t)$. The response of this second system reads $Y(t)$. Calculate the cross-correlation function of $W(t)$ and $Y(t)$ in terms of $h_1(t)$ and $h_2(t)$, and the autocorrelation function of $X(t)$. Assume that $X(t)$ is wide-sense stationary.

4.11 The process (called the signal) $X(t) = \cos(\omega_0 t - \Theta)$, with Θ uniformly distributed on the interval $(0, 2\pi]$, is added to white noise with spectral density $N_0/2$. The sum is applied to the RC network given in Figure 4.2.

(a) Calculate the power spectral densities of the output signal and the output noise.

(b) Calculate the ratio of the mean output signal power and the mean output noise power, the so-called signal-to-noise ratio.

(c) For what value of $\tau_0 = RC$ does this ratio become maximal?

4.12 White noise with spectral density of $N_0/2$ is applied to the input of a linear time-invariant system with the transfer function $H(\omega) = (1 + j\omega\tau_0)^{-2}$. Calculate the power of the output process.

4.13 A differentiating network may be considered as a linear time-invariant system with the transfer function $H(\omega) = j\omega$. If the input is a wide-sense stationary process $X(t)$, then the output process is $dX(t)/dt = \dot{X}(t)$. Show that:

(a) $R_{X\dot{X}}(\tau) = \dfrac{dR_{XX}(\tau)}{d\tau}$

(b) $R_{\dot{X}\dot{X}}(\tau) = -\dfrac{d^2R_{XX}(\tau)}{d\tau^2}$

4.14 To the differentiating network presented in Problem 4.13 a wide-sense stationary process $X(t)$ is applied. The corresponding output process is $Y(t)$.

(a) Are the random variables $X(t)$ and $Y(t)$, both considered at the same fixed time t, orthogonal?

(b) Are these random variables uncorrelated?

4.15 A stochastic process with the power spectral density of

$$S_{XX}(\omega) = \frac{1}{(1 + \omega^2)^2}$$

is applied to a differentiating network.

(a) Find the power spectral density of the output process.

(b) Calculate the power of the derivative of $X(t)$.

4.16 Consider the stochastic process

$$Y(t) = \int_{t-T}^{t+T} X(\xi)\,\mathrm{d}\xi$$

where $X(t)$ is a wide-sense stationary process.

(a) Design a linear time-invariant system that produces the given relation between its input process $X(t)$ and the corresponding output process $Y(t)$.

(b) Express the autocorrelation function of $Y(t)$ in terms of that for $X(t)$.

(c) Express the power spectrum of $Y(t)$ in terms of that for $X(t)$.

4.17 Reconsider Problem 4.16.

(a) Prove that the given integration can be realized by a linear time-invariant filter with an impulse response $\mathrm{rect}[t/(2T)]$.

(b) Fill an array in Matlab with a harmonic function, let us say a cosine. Take, for example, 10 cycles of $1/(2\pi)$ Hz and increments of 0.01 for the time parameter. Plot the function.

(c) Generate a Gaussian noise wave with mean of zero and unit variance, and of the same length as the cosine using the Matlab command `randn`, add this wave to the cosine and plot the result. Note that each time you run the program a different wave is produced.

(d) Program a vector consisting of all ones. Convolve this vector with the cosine plus noise vector and observe the result. Take different lengths, e.g. equivalent to $2T = 0.05$, 0.10 and 0.20. Explain the differences in the different curves.

4.18 In FM detection the white additive noise is converted into noise with spectral density

$$S_{NN}(\omega) = \omega^2$$

and assume that this noise is wide-sense stationary. Suppose that the signal spectrum is

$$S_{XX}(\omega) = \begin{cases} \dfrac{S_0}{2}, & |\omega| \leq W \\ 0, & |\omega| > W \end{cases}$$

and that in the receiver the detected signal is filtered by an ideal lowpass filter of bandwidth W.

(a) Calculate the signal-to-noise ratio.

In audio FM signals so-called pre-emphasis and de-emphasis filtering is applied to improve the signal-to-noise ratio. To that end prior to modulation and transmission the audio baseband signal is filtered by the pre-emphasis filter with the transfer function

$$H(\omega) = 1 + \mathrm{j}\frac{\omega}{W_\mathrm{p}}, \qquad W_\mathrm{p} < W$$

At the receiver side the baseband signal is filtered by the de-emphasis filter such that the spectrum is once more flat and equal to $S_0/2$.

(b) Make a block schematic of the total communication scheme.

(c) Sketch the different signal and noise spectra.

(d) Calculate the improvement factor of the signal-to-noise ratio.

(e) Evaluate the improvement in dB for the practical values: $W/(2\pi) = 15\,\text{kHz}$, $W_p/(2\pi) = 2.1\,\text{kHz}$.

4.19 A so-called nth-order Butterworth filter is defined by the squared value of the amplitude of the transfer function

$$|H(\omega)|^2 = \frac{1}{1 + (\omega/W)^{2n}}$$

where n is an integer, which is called the order of the filter. W is the -3 dB bandwidth in radians per second.

(a) Use Matlab to produce a set of curves that present this squared transfer as a function of frequency; plot the curves on a double logarithmic scale for $n = 1, 2, 3, 4$.

(b) Calculate and evaluate the equivalent noise bandwidth for $n = 1$ and $n = 2$.

4.20 For the transfer function of a bandpass filter it is given that

$$|H(\omega)|^2 = \frac{1}{1 + (\omega - \omega_0)^2} + \frac{1}{1 + (\omega + \omega_0)^2}$$

(a) Use Matlab to plot $|H(\omega)|^2$ for $\omega_0 = 10$.

(b) Calculate the equivalent noise bandwidth of the filter.

(c) Calculate the output noise power when wide-sense stationary noise with spectral density of $N_0/2$ is applied to the input of this filter.

4.21 Consider the so-called Manchester (or split-phase) signalling format defined by

$$p(t) = \begin{cases} 1, & 0 \leq t < T/2 \\ -1, & T/2 \leq t < T \end{cases}$$

where T is the bit time. The data symbols $A[n]$ are selected from the set $\{1, -1\}$ with equal probability and are mutually independent.

(a) Sketch the Manchester coded signal of the sequence 1010111001.

(b) Calculate the power spectral density of this data signal. Use Matlab to plot it.

(c) Discuss the properties of the spectrum in comparison to the polar NRZ signal.

4.22 In the bipolar NRZ signalling format the binary 1's are alternately mapped to $A[n] = +1$ volt and $A[n] = -1$ volt. The binary 0 is mapped to $A[n] = 0$ volt. The bits are selected with equal probability and are mutually independent.

(a) Sketch the bipolar NRZ coded signal of the sequence 1010111001.

(b) Calculate the power spectral density of this data signal. Use Matlab to plot it.

(c) Discuss the properties of the spectrum in comparison to the polar NRZ signal.

4.23 Reconsider Example 4.6. Using Matlab fill in an array of size 256 with a rectangular function of width 50. Apply the FFT procedure to that. Square the resulting array and subsequently apply the IFFT procedure.

(a) Check the FFT result for aliasing.

(b) What in the time domain is the equivalence of squaring in the frequency domain?

(c) Check the IFFT result with respect to your answer to (b).

Now fill another array of size 256 with a rectangular function of width 4 and apply the FFT to it.

(d) Check the result for aliasing.

(e) Multiply the FFT result of the 50 wide pulse by that of the 4 wide pulse and IFFT the multiplication. Is the result what you expected in view of the result from (d)?

4.24 In digital communications a well-known disturbance of the received data symbols is the so-called intersymbol interference (see references [6], [9] and [11]). It is actually the spill-over of the pulse representing a certain bit to the time interval assigned to the adjacent pulses that represent different bits. This disturbance is a consequence of the distortion in the transmission channel. By means of proper filtering, called equalization, the intersymbol interference can be removed or minimized. Assume that each received pulse that represents a bit is sampled once and that the sampled sequence is represented by its z-transform $\tilde{R}(z)$. For an ideal channel, i.e. a channel that does not produce intersymbol interference, we have $\tilde{R}(z) = 1$.

Let us now consider a channel with intersymbol interference and design a discrete-time filter that equalizes the channel. If the z-transform of the filter impulse response is denoted by $\tilde{F}(z)$, then for the equalized pulse the condition $\tilde{R}(z)\,\tilde{F}(z) = 1$ should be satisfied. Therefore, in this problem the sequence $\tilde{R}(z)$ is known and the sequence $\tilde{F}(z)$ has to be solved to satisfy this condition. It appears that the Matlab command `deconv` is not well suited to solving this problem.

(a) Suppose that $\tilde{F}(z)$ comprises three terms. Show that the condition for equalization is equivalent to

$$\begin{bmatrix} r[0] & r[-1] & r[-2] \\ r[1] & r[0] & r[-1] \\ r[2] & r[1] & r[0] \end{bmatrix} \cdot \begin{bmatrix} f[-1] \\ f[0] \\ f[1] \end{bmatrix} = \begin{bmatrix} 0 \\ 1 \\ 0 \end{bmatrix}$$

(b) Consider a received pulse $\tilde{R}(z) = 0.1z + 1 - 0.2z^{-1} + 0.1z^{-2}$. Design the equalizer filter of length 3.

(c) As the quality factor with respect to intersymbol interference we define the 'worst case interference'. It is the sum of the absolute signal samples minus the desired sample value 1. Calculate the output sequence of the equalizer designed in (b) using `conv` and calculate its worst-case interference. Compare this with the unequalized worst-case interference.

(d) Redo the equalizer design for filter lengths 5 and 7, and observe the change in the worst-case interference.

4.25 Find the transfer function and filter structure of the discrete-time system when the following relations exist between the input and output:

(a) $y[n] + 2y[n-1] + 0.5y[n-2] = x[n] - x[n-2]$

(b) $4y[n] + y[n-1] - 2y[n-2] - 2y[n-3] = x[n] + x[n-1] - x[n-2]$

(c) Are the given systems stable?

Hint: use the Matlab command `roots` to compute the roots of polynomials.

4.26 White noise with spectral density of $N_0/2$ is applied to an ideal lowpass filter with bandwidth W.

(a) Calculate the autocorrelation function of the output process. Use Matlab to plot this function.

(b) The output noise is sampled at the time instants $t_n = n\pi/W$ with n integer. What can be remarked with respect to the sample values?

4.27 A discrete-time system has the transfer function $\tilde{H}(z) = 1 + 0.9z^{-1} + 0.7z^{-2}$. To the input of the system the signal with z-transform $\tilde{X}(z) = 0.7 + 0.9z^{-1} + z^{-2}$ is applied. This signal is disturbed by a wide-sense stationary white noise sequence. The autocorrelation sequence of this noise is $R_{NN}[m] = 0.01\,\delta[m]$.

(a) Calculate the signal output sequence.

(b) Calculate the autocorrelation sequence at the output.

(c) Calculate the maximum value of the signal-to-noise ratio. At what moment in time will that occur?

Hint: you can eventually use the Matlab command `conv` to perform the required polynomial multiplications. In this way the solution found using pencil and paper can be checked.

4.28 The transfer function of a discrete-time filter is given by

$$\tilde{H}(z) = \frac{1}{1 - 0.8z^{-1}}$$

(a) Use Matlab's `freqz` to plot the absolute value of the transfer function in the frequency domain.

(b) If the discrete-time system operates at a sampling rate of 1 MHz and a sine wave of 50 kHz and an amplitude of unity is applied to the filter input, compute the power of the corresponding output signal.

(c) A zero mean white Gaussian noise wave is added to the sine wave at the input such that the signal-to-noise ratio amounts to 0 dB. Compute the signal-to-noise ratio at the output.

(d) Use the Matlab command `randn` to generate the noise wave. Design and implement a procedure to test whether the generated noise wave is indeed approximately white noise.

(e) Check the analytical result of (c) by means of proper operations on the waves that are generated by Matlab.

5

Bandpass Processes

Bandpass processes often occur in electrical engineering, mostly as a result of the bandpass filtering of white noise. This is due to the fact that in electrical engineering in general, and specifically in telecommunications, use is made of modulation of signals. These information-carrying signals have to be filtered in systems such as receivers to separate them from other, unwanted signals in order to enhance the quality of the wanted signals and to prepare them for further processing such as detection.

Before dealing with bandpass processes we will present a summary of the description of deterministic bandpass signals.

5.1 DESCRIPTION OF DETERMINISTIC BANDPASS SIGNALS

There are many reasons why signals are modulated. Doing so shifts the spectrum to a certain frequency, so that a bandpass signal results (see Section 3.4). On processing, for example, reception of a telecommunication signal, such signals are bandpass filtered in order to separate them in frequency from other (unwanted) signals and to limit the amount of noise power. We consider signals that consist of a high-frequency carrier modulated in amplitude or phase by a time function that varies much more slowly than the carrier. For instance, amplitude modulation (AM) signals are written as

$$s(t) = A(1 + m(t)) \cos \omega_0 t \tag{5.1}$$

where A is the amplitude of the unmodulated carrier, $m(t)$ is the low-frequency modulating signal and ω_0 is the carrier angular frequency. Note that lower case characters are used here, since in this section we discuss deterministic signals. Assuming that $(1 + m(t))$ is never negative, then $s(t)$ looks like a harmonic signal whose amplitude varies with the modulating signal.

A frequency-modulated signal is written as

$$s(t) = A \cos\left[\omega_0 t + \int_0^t \psi(\tau)\,\mathrm{d}\tau\right] \tag{5.2}$$

Introduction to Random Signals and Noise W. van Etten

The instantaneous angular frequency of this signal is $\omega_0 + \psi(t)$ and is found by differentiating the argument of the cosine with respect to t. In this case the slowly varying function $\psi(t)$ carries the information to be transmitted. The frequency-modulated signal has a constant amplitude, but the zero crossings will change with the modulating signal.

The most general form of a modulated signal is given by

$$s(t) = a(t)\cos[\omega_0 t + \phi(t)] \tag{5.3}$$

In this equation $a(t)$ is the amplitude modulation and $\phi(t)$ the phase modulation, while the derivative $\mathrm{d}\phi(t)/\mathrm{d}t$ represents the frequency modulation of the signal. Expanding the cosine of Equation (5.3) yields

$$s(t) = a(t)[\cos\varphi(t)\cos\omega_0 t - \sin\varphi(t)\sin\omega_0 t] = x(t)\cos\omega_0 t - y(t)\sin\omega_0 t \tag{5.4}$$

with

$$\begin{aligned} x(t) &\triangleq a(t)\cos\varphi(t) \\ y(t) &\triangleq a(t)\sin\varphi(t) \end{aligned} \tag{5.5}$$

The functions $x(t)$ and $y(t)$ are called the quadrature components of the signal. Signal $x(t)$ is called the in-phase component or I-component and $y(t)$ the quadrature or Q-component. They will vary little during one period of the carrier. Combining the quadrature components to produce a complex function will give a representation of the modulated signal in terms of the complex envelope

$$z(t) \triangleq x(t) + \mathrm{j}y(t) = a(t)\exp[\mathrm{j}\varphi(t)] \tag{5.6}$$

When the carrier frequency ω_0 is known, the signal $s(t)$ can unambiguously be recovered from this complex envelope. It is easily verified that

$$s(t) = \mathrm{Re}\{z(t)\exp(j\omega_0 t)\} = \tfrac{1}{2}[z(t)\exp(\mathrm{j}\omega_0 t) + z^*(t)\exp(-\mathrm{j}\omega_0 t)] \tag{5.7}$$

where $\mathrm{Re}\{\cdot\}$ denotes the real part of the quantity in the braces. Together with the carrier frequency ω_0, the signal $z(t)$ constitutes an alternative and complete description of the modulated signal. The expression $z(t)\exp(\mathrm{j}\omega_0 t)$ is called the analytic signal or pre-envelope. The complex function $z(t)$ can be regarded as a phasor in the xy plane. The end of the phasor moves around in the complex plane, while the plane itself rotates with an angular frequency of ω_0 and the signal $s(t)$ is the projection of the rotating phasor on a fixed line. If the movement of the phasor $z(t)$ with respect to the rotating plane is much slower than the speed of rotation of the plane, the signal is quasi-harmonic. The phasor in the complex z plane has been depicted in Figure 5.1.

It is stressed that $z(t)$ is not a physical signal but a mathematically defined auxiliary signal to facilitate the calculations. The name complex envelope suggests that there is a relationship with the envelope of a modulated signal. This envelope is interpreted as

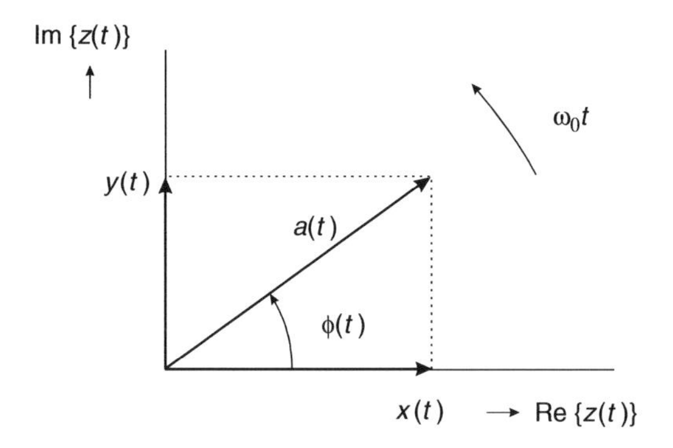

Figure 5.1 The phasor of $s(t)$ in the complex z plane

the instantaneous amplitude of the signal, in this case $a(t)$. Now the relationship is clear, namely

$$\begin{aligned} a(t) &= |z(t)| = \sqrt{x^2(t) + y^2(t)} \\ \varphi(t) &= \arg[z(t)] = \arctan\frac{y(t)}{x(t)} \end{aligned} \tag{5.8}$$

It is concluded that the complex envelope, in contrast to the envelope, not only comprises information about the envelope $a(t)$ but also about the phase $\varphi(t)$.

As far as detection is concerned, the quadrature component $x(t)$ is restored by multiplying $s(t)$ by $\cos\omega_0 t$ and removing the double frequency components with a lowpass filter:

$$s(t)\cos\omega_0 t = \tfrac{1}{2}x(t)(1 + \cos 2\omega_0 t) - \tfrac{1}{2}y(t)\sin 2\omega_0 t \tag{5.9}$$

This multiplication operation can be performed by the modulator scheme presented in Figure 3.5. After lowpass filtering, the signal produced will be $x(t)/2$. The second quadrature component $y(t)$ is restored in a similar way, by multiplying $s(t)$ by $-\sin\omega_0 t$ and using a lowpass filter to remove the double frequency components. A circuit that delivers an output signal that is a function of the amplitude modulation is called a rectifier and such a circuit will always involve a nonlinear operation. The quadratic rectifier is a typical rectifier; it has an output signal in proportion to the square of the envelope. This output is achieved by squaring the signal and reads

$$s^2(t) = \tfrac{1}{2}[x^2(t) + y^2(t)] + \tfrac{1}{2}[x^2(t) - y^2(t)]\cos 2\omega_0 t - x(t)y(t)\sin 2\omega_0 t \tag{5.10}$$

By means of a lowpass filter the frequency terms in the vicinity of $2\omega_0$ are removed, so that the output is proportional to $|z(t)|^2 = a^2(t) = x^2(t) + y^2(t)$. A linear rectifier, which may consist of a diode and a lowpass filter, yields $a(t)$.

A circuit giving an output signal that is proportional to the instantaneous frequency deviation $\varphi'(t)$ is known as a discriminator. Its output is proportional to $\mathrm{d}[\mathrm{Im}\{\ln z(t)\}]/\mathrm{d}t$.

If the signal $s(t)$ comprises a finite amount of energy then its Fourier transform exists. Using Equation (5.7) the spectrum of this signal is found to be

$$\begin{aligned} S(\omega) &= \tfrac{1}{2}\int_{-\infty}^{\infty} [z(t)\exp(\mathrm{j}\omega_0 t) + z^*(t)\exp(-\mathrm{j}\omega_0 t)]\exp(-\mathrm{j}\omega t)\,\mathrm{d}t \\ &= \tfrac{1}{2}[Z(\omega-\omega_0) + Z^*(-\omega-\omega_0)] \end{aligned} \tag{5.11}$$

where $S(\omega)$ and $Z(\omega)$ are the signal spectra (or Fourier transform) of $s(t)$ and $z(t)$, respectively, and * denotes the complex conjugate.

The quadrature components and $z(t)$ vary much more slowly than the carrier and will be baseband signals. The modulus of the spectrum of the signal, $|S(\omega)|$, has two narrow peaks, one at the frequency ω_0 and the other at $-\omega_0$. Consequently, $s(t)$ is called a narrowband signal. The spectrum of Equation (5.11) is Hermitian, i.e. $S(-\omega) = S^*(\omega)$, a condition that is imposed by the fact that $s(t)$ is real.

Quasi-harmonic signals are often filtered by bandpass filters, i.e. filters that pass frequency components in the vicinity of the carrier frequency and attenuate other frequency components. The transfer function of such a filter may be written as

$$H(\omega) = H_1(\omega-\omega_c) + H_1^*(-\omega-\omega_c) \tag{5.12}$$

where the function $H_1(\omega)$ is a lowpass filter; it is called the equivalent baseband transfer function. Equation (5.12) is Hermitian, because $h(t)$, being the impulse response of a physical system, is a real function. However, the equivalent baseband function $H_1(\omega)$ will not be Hermitian in general. Note the similarity of Equations (5.12) and (5.11). The only difference is the carrier frequency ω_0 and the characteristic frequency ω_c. For ω_c, an arbitrary frequency in the passband of $H(\omega)$ may be selected. In Equation (5.7) the characteristic frequency ω_0 need not necessarily be taken as equal to the oscillator frequency. A shift in the characteristic frequency over $\Delta\omega = \omega_1$ merely introduces a factor of $\exp(-\mathrm{j}\omega_1 t)$ in the complex envelope:

$$z(t)\exp(\mathrm{j}\omega_0 t) = [z(t)\exp(-\mathrm{j}\omega_1 t)]\,\exp[\mathrm{j}(\omega_0+\omega_1)t] \tag{5.13}$$

This shift does not change the signal $s(t)$. From this it will be clear that the complex envelope is connected to a specific characteristic frequency; when this frequency changes the complex envelope will change as well. A properly selected characteristic frequency, however, can simplify calculations to a large extent. Therefore, it is important to take the characteristic frequency equal to the oscillator frequency. Moreover, we select the characteristic frequency of the filter equal to that value, i.e. $\omega_c = \omega_0$.

Let us suppose that a modulated signal is applied to the input of a bandpass filter. It appears that using the concepts of the complex envelope and equivalent baseband transfer function the output signal is easily described, as will follow from the sequel. The signal spectra of input and output signals are denoted by $S_i(\omega)$ and $S_o(\omega)$, respectively. It then follows that

$$S_o(\omega) = S_i(\omega)\,H(\omega) \tag{5.14}$$

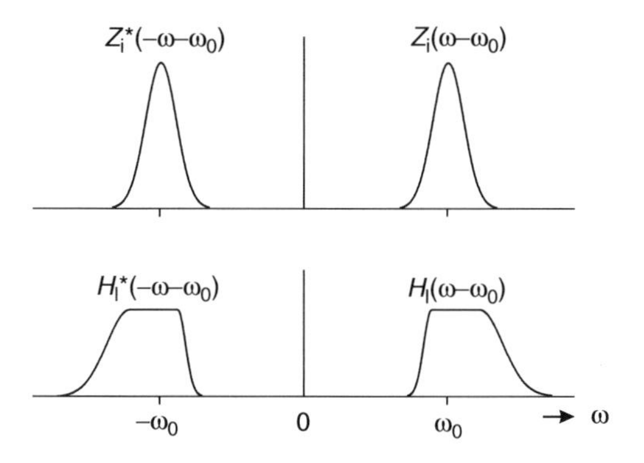

Figure 5.2 A sketch of the cross-terms from the right-hand member of Equation (5.15)

Invoking Equations (5.11) and (5.12) this can be rewritten as

$$\begin{aligned} S_o(\omega) &= \tfrac{1}{2}[Z_o(\omega-\omega_0)+Z_o^*(-\omega-\omega_0)] \\ &= \tfrac{1}{2}[Z_i(\omega-\omega_0)+Z_i^*(-\omega-\omega_0)]\,[H_l(\omega-\omega_0)+H_l^*(-\omega-\omega_0)] \end{aligned} \tag{5.15}$$

where $Z_i(\omega)$ and $Z_o(\omega)$ are the spectra of the complex envelopes of input and output signals, respectively. If it is assumed that the spectrum of the input signal has a bandpass character according to Equations (4.40), (4.41) and (4.42), then the cross-terms $Z_i(\omega-\omega_0)\ H_l^*(-\omega-\omega_0)$ and $Z_i^*(-\omega-\omega_0)\,H_l(\omega-\omega_0)$ vanish (see Figure 5.2). Based on this conclusion Equation (5.15) reduces to

$$Z_o(\omega-\omega_0) = Z_i(\omega-\omega_0)\ H_l(\omega-\omega_0) \tag{5.16}$$

or

$$Z_o(\omega) = Z_i(\omega)\,H_l(\omega) \tag{5.17}$$

Transforming Equation (5.17) to the time domain yields

$$z_o(t) = \int_{-\infty}^{\infty} h_l(\tau) z_i(t-\tau)\,d\tau \tag{5.18}$$

with $z_o(t)$ and $z_i(t)$ the complex envelopes of the input and output signals, respectively. The function $h_l(t)$ is the inverse Fourier transform of $H_l(\omega)$ and represents the complex impulse response of the equivalent baseband system, which is defined by means of $H_l(\omega)$ or the dual description $h_l(t)$. The construction of the equivalent baseband system $H_l(\omega)$ from the actual system is illustrated in Figure 5.3. This construction is quite simple, namely removing the part of the function around $-\omega_0$ and shifting the remaining portion around ω_0 to the baseband, where zero replaces the original position of ω_0. From this figure it is observed that

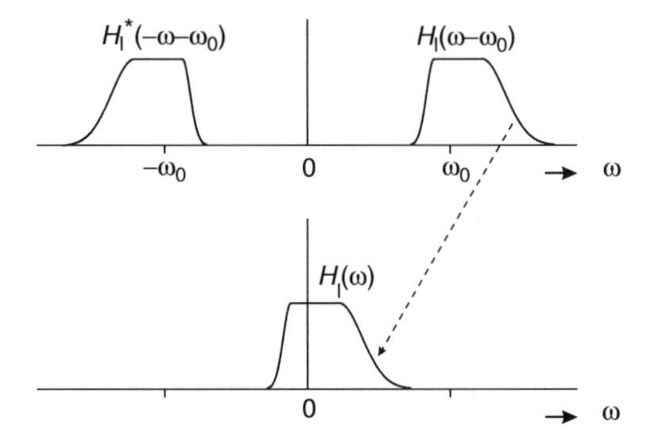

Figure 5.3 Construction of the transfer function of the equivalent baseband system

in general $H_l(\omega)$ is not Hermitian, and consequently $h_l(t)$ will not be a real function. This may not surprise us since it is not the impulse response of a real system but an artificially constructed one that only serves as an intermediate to facilitate calculations.

From Equations (5.17) and (5.18) it is observed that the relation between the output of a bandpass filter and the input (mostly a modulated signal) is quite simple. The signals are completely determined by their complex envelopes and the characteristic frequency ω_0. Using the latter two equations the transmission is reduced to the well-known transmission of a baseband signal. After transforming the signal and the filter transfer function to equivalent baseband quantities the relationship between the output and input is greatly simplified, namely a multiplication in the frequency domain or a convolution in the time domain. Once the complex envelope of the output signal is known, the output signal itself is recovered using Equation (5.7). Of course, using the direct method via Equation (5.14) is always allowed and correct, but many times the method based on the equivalent baseband quantities is simpler, for instance in the case after the bandpass filtering envelope detection is applied. This envelope follows immediately from the complex envelope.

5.2 QUADRATURE COMPONENTS OF BANDPASS PROCESSES

Analogously to modulated deterministic signals (as described in Section 5.1), stochastic bandpass processes may be described by means of their quadrature components. Consider the process

$$N(t) = A(t)\cos[\omega_0 t + \Phi(t)] \tag{5.19}$$

with $A(t)$ and $\Phi(t)$ stochastic processes. The quadrature description of this process is readily found by rewriting this latter equation by applying a basic trigonometric relation

$$\begin{aligned} N(t) &= A(t)\cos\Phi(t)\,\cos\omega_0 t - A(t)\sin\Phi(t)\,\sin\omega_0 t \\ &= X(t)\cos\omega_0 t - Y(t)\sin\omega_0 t \end{aligned} \tag{5.20}$$

In this expression the quadrature components are defined as

$$\begin{aligned} X(t) &\triangleq A(t)\cos\Phi(t) \\ Y(t) &\triangleq A(t)\sin\Phi(t) \end{aligned} \tag{5.21}$$

From these equations the processes describing the amplitude and phase of $N(t)$ are easily recovered:

$$\begin{aligned} A(t) &\triangleq \sqrt{X^2(t)+Y^2(t)} \\ \Phi(t) &\triangleq \arctan\left[\frac{Y(t)}{X(t)}\right] \end{aligned} \tag{5.22}$$

The processes $X(t)$ and $Y(t)$ are stochastic lowpass processes (or baseband processes), while Equation (5.20) presents a general description of bandpass processes. In the sequel we will derive relations between these lowpass processes $X(t)$ and $Y(t)$, and the lowpass processes on the one hand and the bandpass process $N(t)$ on the other. Those relations refer to mean values, correlation functions, spectra, etc., i.e. characteristics that have been introduced in previous chapters.

Once again we assume that the process $N(t)$ is wide-sense stationary with a mean value of zero. A few interesting properties can be stated for such a bandpass process. The properties are given below and proved subsequently.

Properties of bandpass processes

If $N(t)$ is a wide-sense stationary bandpass process with mean value zero and quadrature components $X(t)$ and $Y(t)$, then $X(t)$ and $Y(t)$ have the following properties:

1. $X(t)$ and $Y(t)$are jointly wide-sense stationary. (5.23)
2. $\mathrm{E}[X(t)] = \mathrm{E}[Y(t)] = 0$ (5.24)
3. $\mathrm{E}[X^2(t)] = \mathrm{E}[Y^2(t)] = \mathrm{E}[N^2(t)]$ (5.25)
4. $R_{XX}(\tau) = R_{YY}(\tau)$ (5.26)
5. $R_{YX}(\tau) = -R_{XY}(\tau)$ (5.27)
6. $R_{XY}(0) = R_{YX}(0) = 0$ (5.28)
7. $S_{YY}(\omega) = S_{XX}(\omega)$ (5.29)
8. $S_{XX}(\omega) = \mathrm{Lp}\{S_{NN}(\omega-\omega_0) + S_{NN}(\omega+\omega_0)\}$ (5.30)
9. $S_{XY}(\omega) = \mathrm{j}\,\mathrm{Lp}\{S_{NN}(\omega-\omega_0) - S_{NN}(\omega+\omega_0)\}$ (5.31)
10. $S_{YX}(\omega) = -S_{XY}(\omega)$ (5.32)

In Equations (5.30) and (5.31), $\mathrm{Lp}\{\cdot\}$ denotes the lowpass part of the expression in the braces.

Proofs of the properties:

Here we shall briefly prove the properties listed above. A few of them will immediately be clear. Expression (5.29) follows directly from Equation (5.26), and Equation (5.32) from Equation (5.27). Moreover, using Equation (2.48), Equation (5.28) is a consequence of Equation (5.27).

Invoking the definition of the autocorrelation function, it follows after some manipulation from Equation (5.20) that

$$\begin{aligned}R_{NN}(t,t+\tau) = \tfrac{1}{2}\{&[R_{XX}(t,t+\tau) + R_{YY}(t,t+\tau)]\cos\omega_0\tau \\ &- [R_{XY}(t,t+\tau) - R_{YX}(t,t+\tau)]\sin\omega_0\tau \\ &+ [R_{XX}(t,t+\tau) - R_{YY}(t,t+\tau)]\cos\omega_0(2t+\tau) \\ &- [R_{XY}(t,t+\tau) + R_{YX}(t,t+\tau)]\sin\omega_0(2t+\tau)\}\end{aligned} \tag{5.33}$$

Since we assumed that $N(t)$ is a wide-sense stationary process, Equation (5.33) has to be independent of t. Then, from the last term of Equation (5.33) it is concluded that

$$R_{XY}(t,t+\tau) = -R_{YX}(t,t+\tau) \tag{5.34}$$

and from the last but one term of Equation (5.33)

$$R_{XX}(t,t+\tau) = R_{YY}(t,t+\tau) \tag{5.35}$$

Using these results it follows from the first two terms of Equation (5.33) that

$$R_{XX}(t,t+\tau) = R_{XX}(\tau) = R_{YY}(\tau) \tag{5.36}$$

and

$$R_{XY}(t,t+\tau) = R_{XY}(\tau) = -R_{YX}(\tau) \tag{5.37}$$

thereby establishing properties 4 and 5. Equation (5.33) can now be rewritten as

$$R_{NN}(\tau) = R_{XX}(\tau)\cos\omega_0\tau - R_{XY}(\tau)\sin\omega_0\tau \tag{5.38}$$

If we substitute $\tau = 0$ in this expression and use Equation (5.26), property 3 follows.

The expected value of $N(t)$ reads

$$\mathrm{E}[N(t)] = \mathrm{E}[X(t)]\cos\omega_0 t - \mathrm{E}[Y(t)]\sin\omega_0 t = 0 \tag{5.39}$$

This equation is satisfied only if

$$\mathrm{E}[X(t)] = \mathrm{E}[Y(t)] = 0 \tag{5.40}$$

so that now property 2 has been established. However, this means that now property 1 has been proved as well, since the mean values of $X(t)$ and $Y(t)$ are independent of t

(property 2) and also the autocorrelation and cross-correlation functions Equations (5.36) and (5.37).

By transforming Equation (5.38) to the frequency domain we arrive at

$$S_{NN}(\omega) = \tfrac{1}{2}[S_{XX}(\omega-\omega_0) + S_{XX}(\omega+\omega_0)] + \tfrac{1}{2}\mathrm{j}[S_{XY}(\omega-\omega_0) - S_{XY}(\omega+\omega_0)] \qquad (5.41)$$

and using this expression gives

$$S_{NN}(\omega-\omega_0) = \tfrac{1}{2}[S_{XX}(\omega-2\omega_0) + S_{XX}(\omega)] + \tfrac{1}{2}\mathrm{j}[S_{XY}(\omega-2\omega_0) - S_{XY}(\omega)] \qquad (5.42)$$

$$S_{NN}(\omega+\omega_0) = \tfrac{1}{2}[S_{XX}(\omega) + S_{XX}(\omega+2\omega_0)] + \tfrac{1}{2}\mathrm{j}[S_{XY}(\omega) - S_{XY}(\omega+2\omega_0)] \qquad (5.43)$$

Adding Equations (5.42) and (5.43) produces property 8. Subtracting Equation (5.43) from Equation (5.42) produces property 9.

Some of those properties are very peculiar; namely the quadrature processes $X(t)$ and $Y(t)$ both have a mean value of zero, are wide-sense stationary, have identical autocorrelation functions and as a consequence have the same spectrum. The processes $X(t)$ and $Y(t)$ comprise the same amount of power and this amount equals that of the original bandpass process $N(t)$ (property 3). At first sight this property may surprise us, but after a closer inspection it is recalled that $X(t)$ and $Y(t)$ are the quadrature processes of $N(t)$, and then the property is obvious. Finally, at each moment of time t, the random variables $X(t)$ and $Y(t)$ are orthogonal (property 6).

When the spectrum of $N(t)$ is symmetrical about ω_0, the stochastic processes $X(t)$ and $Y(t)$ are orthogonal processes (this follows from property 9 and will be further explained by the next example). In the situation at hand the cross-power spectral density is identical to zero. If, moreover, the processes $X(t)$ and $Y(t)$ are Gaussian, then they are also independent.

Example 5.1:

In Figure 5.4(a) an example is depicted of a spectrum of a bandpass process. In this figure the position of the characteristic frequency ω_0 is clearly indicated. On the positive part of the x axis the spectrum covers the region from $\omega_0 - W_1$ to $\omega_0 + W_2$ and on the negative part of the x axis from $-\omega_0 - W_2$ to $-\omega_0 + W_1$. Therefore, the bandwidth of the process is $W = W_1 + W_2$. For a bandpass process the requirement $W_1 < \omega_0$ has to be satisfied.

In Figure 5.4(b) the spectrum $S_{NN}(\omega - \omega_0)$ is presented; this spectrum is obtained by shifting the spectrum given in Figure 5.4(a) by ω_0 to the right. Similarly, the spectrum $S_{NN}(\omega + \omega_0)$ is given in Figure 5.4(c), and this figure is yielded by shifting the spectrum of Figure 5.4(a) by ω_0 to the left. From Figures 5.4(b) and (c) the spectra of the quadrature components can be constructed using the relations of Equations (5.30) and (5.31). By adding the spectra of Figure 5.4(b) and Figure 5.4(c) the spectra $S_{XX}(\omega)$ and $S_{YY}(\omega)$ are found (see Figure 5.4(d)). Next, by subtracting the spectra of Figure 5.4(b) and Figure 5.4(c) we arrive at $-\mathrm{j}S_{XY}(\omega) = \mathrm{j}S_{YX}(\omega)$ (see Figure 5.4(e)). When adding and subtracting as described above, those parts of the spectra that are concentrated about $2\omega_0$ and $-2\omega_0$ have to be ignored. This is in accordance with Equations (5.30) and (5.31). These equations include a lowpass filtering after addition and subtraction, respectively.

□

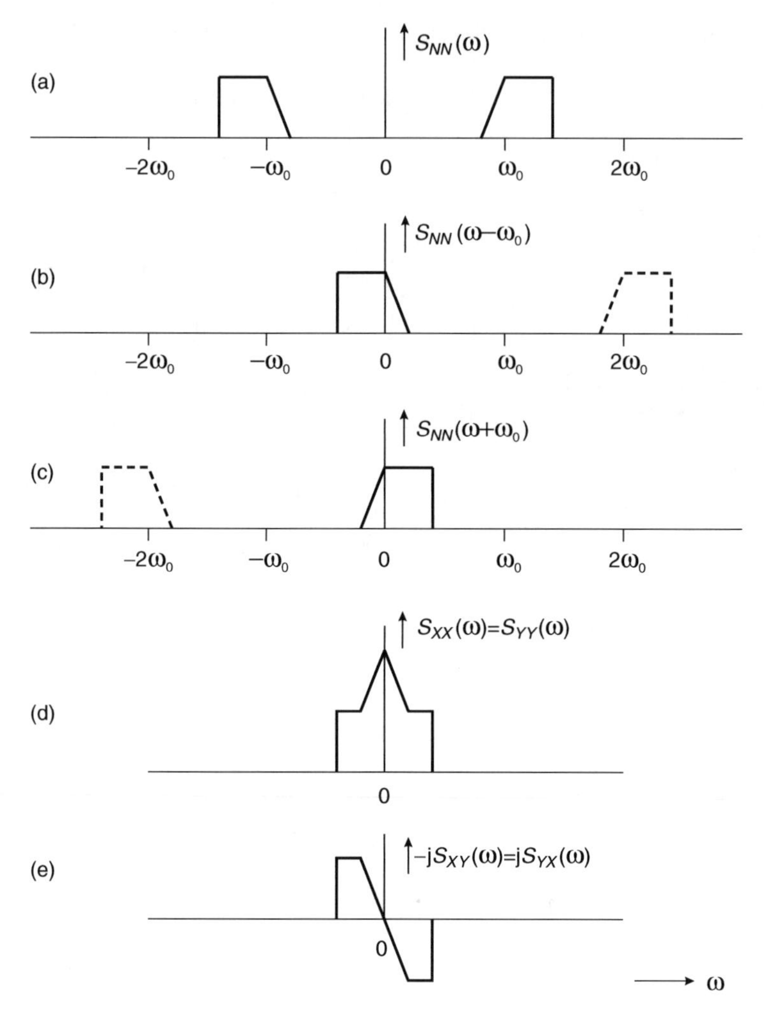

Figure 5.4 Example of a bandpass process and construction of the related quadrature processes

From a careful look at this example it becomes clear that care should be taken when using properties 8 and 9. This is a consequence of the operation $\text{Lp}\{\cdot\}$, which is not always unambiguous. Although there is a different mathematical approach to avoid this, we will not go into the details here. It will be clear that no problems will arise in the case of narrowband bandpass processes. In many cases a bandpass process results from bandpass filtering of white noise. Then the spectrum of the bandpass noise is determined by the equation (see Theorem 7)

$$S_{NN}(\omega) = S_{II}(\omega_0)|H(\omega)|^2 \tag{5.44}$$

where $S_{II}(\omega_0)$ is the spectral density of the input noise and $H(\omega)$ the transfer function of the bandpass filter.

It should be stressed here that the quadrature processes $X(t)$ and $Y(t)$ are not uniquely determined; namely it follows from Equation (5.20) and Figure 5.4 that these processes are, among others, determined by the choice of the characteristic frequency ω_0.

Finally, we will derive the relation between the spectrum of the complex envelope and the spectra of the quadrature components. The complex envelope of a stochastic bandpass process is a complex stochastic process defined by

$$Z(t) \triangleq X(t) + \mathrm{j}Y(t) \tag{5.45}$$

Using Equations (5.26), (5.27) and (2.87) we find that

$$R_{ZZ}(\tau) = 2[R_{XX}(\tau) + \mathrm{j}R_{XY}(\tau)] \tag{5.46}$$

and consequently

$$S_{ZZ}(\omega) = 2[S_{XX}(\omega) + \mathrm{j}S_{XY}(\omega)] \tag{5.47}$$

If $S_{NN}(\omega)$ is symmetrical about ω_0, then $S_{XY}(\omega) = 0$ and the spectrum of the complex envelope reads

$$S_{ZZ}(\omega) = 2S_{XX}(\omega) \tag{5.48}$$

It has been observed that the complex envelope is of importance when establishing the envelope of a bandpass signal or bandpass noise. Equation (5.48) is needed when analysing the envelope detection of (amplitude) modulated signals disturbed by noise.

5.3 PROBABILITY DENSITY FUNCTIONS OF THE ENVELOPE AND PHASE OF BANDPASS NOISE

As mentioned in Section 5.2, in practice we often meet a situation that can be modelled by white Gaussian noise that is bandpass filtered. Linear filtering of Gaussian noise in turn produces Gaussian distributed noise at the filter output, and of course this holds for the special case of bandpass filtered Gaussian noise. Moreover, we conclude that the quadrature components $X(t)$ and $Y(t)$ of Gaussian bandpass noise have Gaussian distributions as well. This is reasoned as follows. Consider the description of bandpass noise in accordance with Equation (5.20). For a certain fixed value of t, let us say t_1, the random variable $N(t_1)$ is constituted from a linear combination of the two random variables $X(t_1)$ and $Y(t_1)$, namely

$$N(t_1) = X(t_1)\cos\omega_0 t_1 - Y(t_1)\sin\omega_0 t_1 \tag{5.49}$$

The result $N(t_1)$ can only be a Gaussian random variable if the two constituting random variables $X(t_1)$ and $Y(t_1)$ show Gaussian distributions as well. From Equations (5.24) and (5.25) we saw that these random variables have a mean value of zero and the same variance σ^2, so they are identically distributed. As $X(t_1)$ and $Y(t_1)$ are Gaussian and orthogonal (see Equation (5.28)), they are independent. In Section 5.2 it was concluded that in case the bandpass filter, and thus the filtered spectrum, is symmetrical about the characteristic frequency, the cross-correlation between the quadrature components is zero and consequently the quadrature components are independent.

In a number of applications the problem arises about the probability density functions of the envelope and phase of a bandpass filtered white Gaussian noise process. In ASK or FSK systems, for instance, in addition to this noise there is still a sine or cosine wave of frequency within the passband of the filter. When ASK or FSK signals are detected incoherently (which is preferred for the sake of simplicity) and we want to calculate the performance of these systems, the probability density function of the envelope of cosine (or sine) plus noise is needed. For coherent detection we need to have knowledge about the phase as well.

We are therefore looking for the probability density functions of the envelope and phase of the process

$$N(t) + C\cos\omega_0 t = [X(t) + C]\cos\omega_0 t - Y(t)\sin\omega_0 t \tag{5.50}$$

where $C\cos\omega_0 t$ is the information signal and quadrature components $Y(t)$ and

$$\Xi(t) \triangleq X(t) + C \tag{5.51}$$

These quadrature components describe the process in rectangular coordinates, while we need a description on the basis of polar coordinates. When the processes for amplitude and phase are denoted by $A(t)$ and $\Phi(t)$, respectively, it follows that

$$A(t) = \sqrt{\Xi^2(t) + Y^2(t)} \tag{5.52}$$

$$\Phi(t) = \arctan\left[\frac{Y(t)}{\Xi(t)}\right] \tag{5.53}$$

The conversion from rectangular to polar coordinates is depicted in Figure 5.5. The probability that the outcome of a realization (ξ, y) of the random variables (Ξ, Y) lies in the region $(a, a + \mathrm{d}a, \phi, \phi + \mathrm{d}\phi)$ is found by the coordinates transformation

$$\xi = a\cos\phi, \quad y = a\sin\phi, \quad \mathrm{d}\xi\,\mathrm{d}y = a\,\mathrm{d}a\,\mathrm{d}\phi \tag{5.54}$$

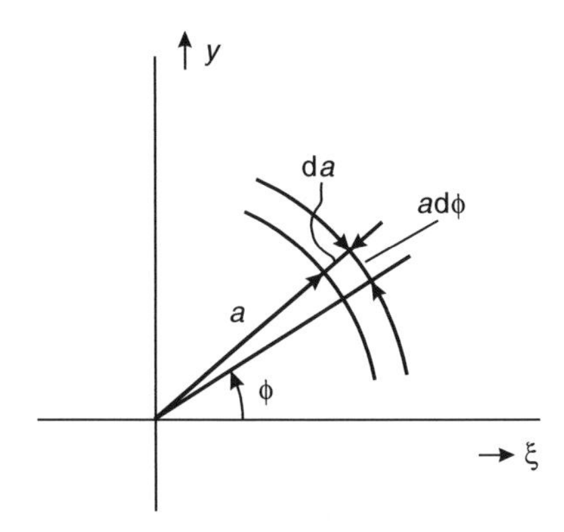

Figure 5.5 Conversion from rectangular to polar coordinates

and is written as

$$f_{XY}(x,y)\,\mathrm{d}x\,\mathrm{d}y = \frac{1}{2\pi\sigma^2}\exp\left[-\frac{(\xi - C)^2 + y^2}{2\sigma^2}\right]\mathrm{d}x\;\mathrm{d}y$$
$$= \frac{1}{2\pi\sigma^2}\exp\left[-\frac{(a\cos\phi - C)^2 + a^2\sin^2\phi}{2\sigma^2}\right]a\;\mathrm{d}a\;\mathrm{d}\phi \tag{5.55}$$

From this the joint probability density function follows as

$$f_{A\Phi}(a,\phi) = \frac{1}{2\pi\sigma^2}a\exp\left[-\frac{(a\cos\phi - C)^2 + a^2\sin^2\phi}{2\sigma^2}\right] \tag{5.56}$$

The marginal probability density function of A is found by integration of this function with respect to ϕ:

$$f_A(a) = \frac{1}{2\pi\sigma^2}\int_0^{2\pi}\exp\left[-\frac{(a\cos\phi - C)^2 + a^2\sin^2\phi}{2\sigma^2}\right]a\;\mathrm{d}a\;\mathrm{d}\phi$$
$$= \frac{1}{2\pi\sigma^2}a\exp\left(-\frac{a^2 + C^2}{2\sigma^2}\right)\int_0^{2\pi}\exp\left(\frac{Ca\cos\phi}{\sigma^2}\right)\mathrm{d}\phi,\quad a \geq 0 \tag{5.57}$$

In this equation the integral cannot be expressed in a closed form but is closely related to the modified Bessel function of the first kind and zero order. This Bessel function can be defined by

$$\mathrm{I}_0(x) \triangleq \frac{1}{2\pi}\int_0^{2\pi}\exp(x\cos\phi)\,\mathrm{d}\phi \tag{5.58}$$

Using this definition the probability density function of A is written as

$$f_A(a) = \frac{a}{\sigma^2}\exp\left(-\frac{a^2 + C^2}{2\sigma^2}\right)\mathrm{I}_0\left(\frac{Ca}{\sigma^2}\right),\quad a \geq 0 \tag{5.59}$$

This expression presents the probability density function for the general case of bandpass filtered white Gaussian noise added to an harmonic signal $C\cos\omega_0 t$ that lies in the passband. This distribution is called the Rice distribution. In Figure 5.6 a few Rice probability density functions are given for several values of C and for all curves where $\sigma = 1$. A special case can be distinguished, namely when $C = 0$, where the signal only consists of bandpass noise since the amplitude of the harmonic signal is set to zero. Then the probability density function is

$$f_A(a) = \frac{a}{\sigma^2}\exp\left(-\frac{a^2}{2\sigma^2}\right),\quad a \geq 0 \tag{5.60}$$

This latter probability density function corresponds to the so-called Rayleigh-distribution. Its graph is presented in Figure 5.6 as well.

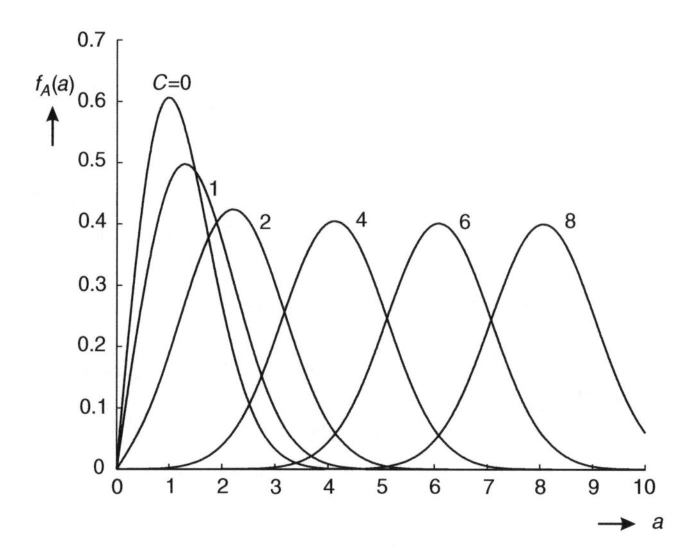

Figure 5.6 Rayleigh distribution ($C = 0$) and Rice distribution for several values of $C \neq 0$ and for $\sigma = 1$

Next we will calculate the probability density function of the phase. For that purpose we integrate the joint probability density function of Equation (5.56) with respect to a. The marginal probability density function of Φ is found as

$$\begin{aligned} f_\Phi(\phi) &= \frac{1}{2\pi\sigma^2}\int_{a=0}^{\infty} a\exp\left(-\frac{a^2 - 2Ca\cos\phi + C^2\cos^2\phi + C^2\sin^2\phi}{2\sigma^2}\right)\mathrm{d}a \\ &= \frac{1}{2\pi\sigma^2}\int_{a=0}^{\infty} a\exp\left[-\frac{(a - C\cos\phi)^2 + C^2\sin^2\phi}{2\sigma^2}\right]\mathrm{d}a \\ &= \frac{1}{2\pi\sigma^2}\exp\left[-\frac{C^2\sin^2\phi}{2\sigma^2}\right]\int_{a=0}^{\infty} a\exp\left[-\frac{(a - C\cos\phi)^2}{2\sigma^2}\right]\mathrm{d}a \end{aligned} \tag{5.61}$$

By means of the change of the integration variable

$$u \triangleq \frac{a - C\cos\phi}{\sigma} \tag{5.62}$$

we proceed to obtain

$$\begin{aligned} f_\Phi(\phi) &= \frac{1}{2\pi\sigma^2}\exp\left(-\frac{C^2\sin^2\phi}{2\sigma^2}\right)\int_{-\frac{C\cos\phi}{\sigma}}^{\infty}(u\sigma + C\cos\phi)\exp\left(-\frac{u^2}{2}\right)\sigma\,\mathrm{d}u \\ &= \frac{1}{2\pi}\exp\left(-\frac{C^2\sin^2\phi}{2\sigma^2}\right)\left[\int_{-\frac{C\cos\phi}{\sigma}}^{\infty}\exp\left(-\frac{u^2}{2}\right)\mathrm{d}\frac{u^2}{2} + \frac{C\cos\phi}{\sigma}\int_{-\frac{C\cos\phi}{\sigma}}^{\infty}\exp\left(-\frac{u^2}{2}\right)\mathrm{d}u\right] \\ &= \frac{1}{2\pi}\exp\left(-\frac{C^2}{2\sigma^2}\right) + \frac{1}{\sigma\sqrt{2\pi}}C\cos\phi\,\exp\left(-\frac{C^2\sin^2\phi}{2\sigma^2}\right)\left[1 - \mathrm{Q}\left(\frac{C\cos\phi}{\sigma}\right)\right], \quad |\phi| < \pi \end{aligned} \tag{5.63}$$

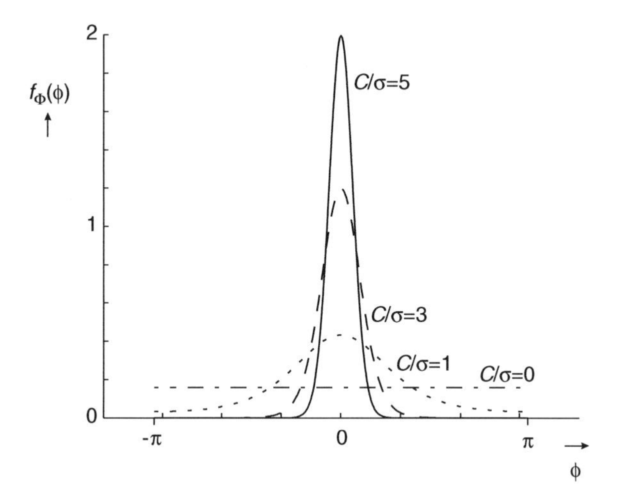

Figure 5.7 Probability density function for the phase of a harmonic signal plus bandpass noise

where the Q function is the well-known Gaussian integral function as defined in Appendix F. The phase probability density function is depicted in Figure 5.7. This unattractive phase equation reduces drastically when $C = 0$ is inserted:

$$f_\Phi(\phi) = \frac{1}{2\pi}, \quad |\phi| < \pi \tag{5.64}$$

Thus, it is concluded that in the case of the Rayleigh distribution the phase has a uniform probability density function. Moreover, in this case the joint probability density function of Equation (5.56) becomes independent of ϕ, so that from a formal point of view this function can be factored according to $f_{A\Phi}(a, \phi) = f_A(a)f_\Phi(\phi)$. Thus, for the Rayleigh distribution, the envelope and phase are independent. This is certainly not true for the Rice distribution ($C \neq 0$). Then the expression of Equation (5.63) can, however, be simplified if the power of the sinusoidal signal is much larger than the noise variance, i.e. $C \gg \sigma$:

$$f_\Phi(\phi) \approx \frac{C \cos \phi}{\sqrt{2\pi}\, \sigma} \exp\left(-\frac{C^2 \sin^2 \phi}{2\sigma^2}\right), \quad |\phi| < \frac{\pi}{2} \tag{5.65}$$

which is approximated by a Gaussian distribution for small values of ϕ, and $\mathrm{E}[\Phi] \approx 0$ and $\mathrm{E}[\Phi^2] \approx \sigma^2$.

5.4 MEASUREMENT OF SPECTRA

5.4.1 The Spectrum Analyser

The oscilloscope is the most well-known measuring instrument used to show and record how a signal behaves as a function of time. However, when dealing with stochastic signals this

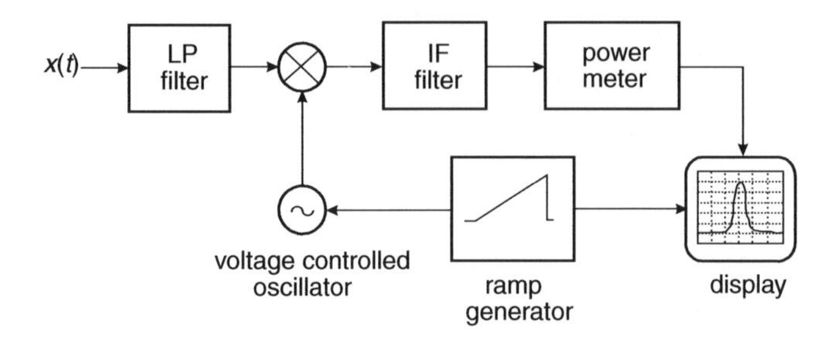

Figure 5.8 Basic scheme of the superheterodyne spectrum analyser

will provide little information due to random fluctuations with time. Measurements in the frequency domain, especially the power spectrum as described in this section, is much more meaningful in these situations. Here we discuss and analyse the mode of operation of the spectrum analyser.

The simplest way to measure a power spectrum is to apply the stochastic signal to a narrowband bandpass filter, to square the filter output and to record the result averaged over a sufficiently long period of time. By tuning the central frequency of the bandpass filter and recording the squared output as a function of frequency an approximation of the spectrum is achieved. However, it is very difficult to tune the frequency of such a filter over a broad range. Even more difficult is to guarantee the properties (such as a constant bandwidth) over a broad frequency range. Therefore, in modern spectrum analysers a different technique based on superheterodyning is used, which is discussed in the sequel.

In Figure 5.8 the basic scheme of this spectrum analyser is presented. First the signal is applied to a lowpass filter (LP), after which it is multiplied by the signal of a local oscillator, which is a voltage controlled oscillator. The angular frequency of this oscillator is denoted by ω_L. Due to this multiplication the spectrum of the input signal is shifted by the value of the frequency; this shifting process has been explained in Section 3.4. Next, the shifted spectrum is filtered by a narrowband bandpass filter (IF), of which the central frequency is ω_{IF}. By keeping the filter parameters (central frequency and bandwidth) fixed but tuning the local oscillator frequency for different local oscillator frequencies, different portions of the power spectrum are filtered. This has the same effect as tuning the filter, which was suggested in the previous paragraph. The advantage of this so-called superheterodyne technique is that the way the filter filters out small portions of the spectrum is the same for all frequencies. The output of the filter is applied to a power meter. The tuning of the voltage controlled oscillator is determined by a generator that generates a linear increasing voltage as a function of time (ramp function), while the frequency is proportional to this voltage. This ramp function is used as the horizontal coordinate in the display unit, whereas the vertical coordinate is proportional to the output of the power meter. The shift of the spectrum caused by the multiplication is depicted in Figure 5.9. Looking at this figure and using Equation (3.38), we can conclude that the power after the IF filter and at a local oscillator frequency of ω_L is written as

$$\begin{aligned} P_Y(\omega_L) &= \frac{1}{2\pi}\int_{-\infty}^{\infty} \frac{A_0^2}{4}[S_{XX}(\omega-\omega_L)+S_{XX}(\omega+\omega_L)]\,|H(\omega)|^2\,d\omega \\ &= 2\frac{1}{2\pi}\frac{A_0^2}{4}\int_0^{\infty} S_{XX}(\omega-\omega_L)\,|H(\omega)|^2\,d\omega \end{aligned} \tag{5.66}$$

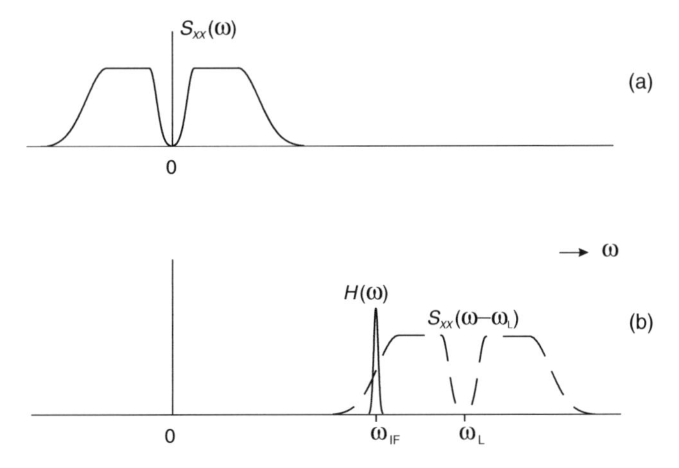

Figure 5.9 (a) Spectrum to be measured; (b) spectrum after mixing and transfer function of the IF filter

where $H(\omega)$ is the transfer function of the IF filter and A_0 is the amplitude of the local oscillator signal. This shows a second advantage of the superheterodyne method; namely the recorded signal is boosted by the amplitude of the local oscillator signal, making it less vulnerable to noise during processing. If the spectrum of the input signal is supposed to be constant over the passband of the IF filter, then according to Equation (4.52) the measured power reads

$$P_Y(\omega_L) \approx \frac{A_0^2}{4\pi} S_{XX}(\omega_{IF} - \omega_L) \left| H(\omega_{IF}) \right|^2 W_N \tag{5.67}$$

Thus, from the power indication of the instrument it appears that the spectrum at the frequency $\omega_{IF} - \omega_L$ is approximated by

$$S_{XX}(\omega_{IF} - \omega_L) \approx \frac{4\pi}{A_0^2 W_N \left| H(\omega_{IF}) \right|^2} P_Y(\omega_L) \tag{5.68}$$

Here ω_{IF} is a fixed frequency that is determined by the design of the spectrum analyser and ω_L is determined by the momentary tuning of the local oscillator in the instrument. Both values are known to the instrument and thus the frequency $\omega_{IF} - \omega_L$ at which the spectral density is measured is known. Furthermore, the fraction in the right-hand member of this equation is determined by calibration and inserted in the instrument, so that reliable data can be measured.

In order to gain an insight into the measured spectra a few design parameters have to be considered in more detail. The local oscillator frequency is chosen such that its value is larger than the IF frequency of the filter. From Equation (5.68) it becomes evident that in that case the frequency at which the spectrum is measured is actually negative. This is no problem at all, as the spectrum is an even function of ω; in other words, that frequency may be replaced by its equal magnitude positive counterpart $\omega_L - \omega_{IF}$. Furthermore, the multiplication will in practice be realized by means of a so-called 'mixer', a non-linear device (e.g. a diode). This device reproduces, besides the sum and difference frequencies, the original frequencies. In order to prevent these components from entering the bandpass filter, the IF frequency has to be higher than the highest component of the signal. The scheme shows a lowpass filter at the input of the analyser; this filter limits the frequency range of the signal that is applied to the mixer in order to make sure that the mixed signal comprises no direct components within the passband range of the IF filter. Moreover, the mixer also produces a signal of the difference frequency, by which the spectrum in Figure 5.9 shifts to the left as well. The input filter also has to prevent these components from passing into the IF filter.

In modern spectrum analysers the signal to be measured is digitized and applied to a digital built-in signal processor that generates the ramp signal (see Figure 5.8). The output of the processor controls the display. This processor is also able to perform all kinds of arithmetical and mathematical operations on the measurement results, such as calculating the maximum value, averaging over many recordings, performing logarithmic scale conversion for decibel presentation, bandwidth indication, level indication, etc.

5.4.2 Measurement of the Quadrature Components

The measurement of the quadrature components of a stochastic process is done by synchronous demodulation, as described in Section 3.4. The bandpass process $N(t)$ is multiplied by a cosine wave as well as by a sine wave, as shown in Figure 5.10. Multiplying by a cosine as is done in the upper arm gives

$$\begin{aligned} 2N(t)\cos(\omega_0 t) &= 2X(t)\cos^2(\omega_0 t) - 2Y(t)\sin(\omega_0 t)\cos(\omega_0 t) \\ &= X(t) + X(t)\cos(2\omega_0 t) - Y(t)\sin(2\omega_0 t) \end{aligned} \tag{5.69}$$

When the bandwidth of the quadrature components is smaller than the central frequency ω_0, the double frequency terms are removed by the lowpass filter, so that the component $X(t)$ is left at the output.

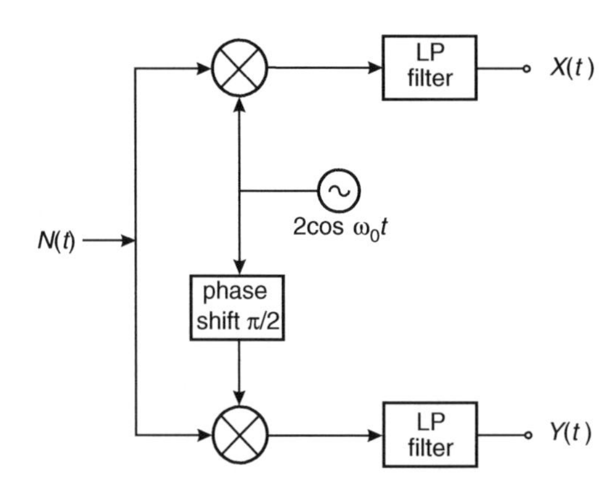

Figure 5.10 Scheme for measuring the quadrature components $X(t)$ and $Y(t)$ of a bandpass process

Multiplying by the sine wave in the lower arm yields

$$\begin{aligned}-2N(t)\sin(\omega_0 t) &= -2X(t)\cos(\omega_0 t)\sin(\omega_0 t) + Y(t)\sin^2(\omega_0 t)\\ &= X(t)\sin(2\omega_0 t) + Y(t) - Y(t)\cos(2\omega_0 t)\end{aligned} \tag{5.70}$$

and removing the double frequency terms using the lowpass filter produces the $Y(t)$ component. If needed, the corresponding output spectra can be measured using the method described in the previous section.

When the cross-power spectrum $S_{XY}(\omega)$ is required we have to cross-correlate the outputs and Fourier transform the cross-correlation function.

5.5 SAMPLING OF BANDPASS PROCESSES

In Section 3.5, by means of Theorem 5 it was shown that a band-limited lowpass stochastic process can be perfectly reconstructed (in the mean-squared-error sense) if the sampling rate is at least equal to the highest frequency component of the spectrum. Applying this theorem to bandpass processes would result in very high sampling rates. Intuitively, one would expect that for these processes the sampling rate should be related to bandwidth, rather than the highest frequency. This is confirmed in the next sections.

Looking for a minimum sampling rate is motivated by the fact that one wants to process as little data as possible using the digital processor, owing to its limited processing and memory capacity. Once the data are digitized, filtering is performed as described in the foregoing sections.

5.5.1 Conversion to Baseband

The simplest way to approach sampling of bandpass processes is to convert them to baseband. This can be performed by means of the scheme given in Figure 5.10 and described in Subsection 5.4.2. Then it can be proved that the sampling rate has to be at least twice the bandwidth of the bandpass process. This is explained as follows. If the demodulation frequency ω_0 in Figure 5.10 is selected in the centre of the bandpass spectrum, then the spectral width of both $X(t)$ and $Y(t)$ is $W/2$, where W is the spectral width of the bandpass process. Each of the quadrature components requires a sampling rate of at least $W/(2\pi)$ (twice its highest frequency component). Therefore, in total W/π samples per second are required to determine the bandpass process $N(t)$; this is the same amount as for a lowpass process with the same bandwidth W as the bandpass process.

The original bandpass process can be reconstructed from its quadrature components $X(t)$ and $Y(t)$ by remodulating them by respectively a cosine and a sine wave and subtracting the modulated signals according to Equation (5.20).

5.5.2 Direct Sampling

At first glance we would conclude that direct sampling of the bandpass process leads to unnecessarily high sampling rates. However, in this subsection we will show that direct

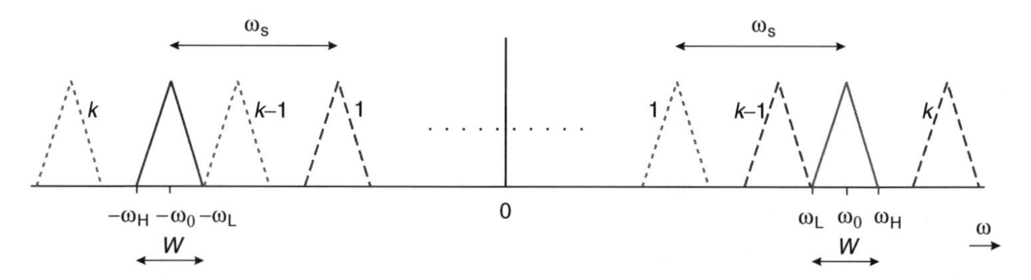

Figure 5.11 Direct sampling of the passband process

sampling is possible at relatively low rates. This is understood when looking at Figure 5.11. The solid lines represent the spectrum of the passband process. When applying ideal sampling, this spectrum is reproduced infinitely many times, where the distance in frequency between adjacent copies is $\omega_s = 2\pi/T_s$. The spectrum of the sampled passband process can therefore be denoted as

$$S_{SS}(\omega) = \sum_{k=-\infty}^{k=\infty} S_{NN}\left(\omega - k\frac{2\pi}{T_s}\right) \tag{5.71}$$

The spectrum $S_{NN}(\omega)$ comprises two portions, one centred around ω_0 and another centred around $-\omega_0$. In Figure 5.11 the shifted versions of $S_{NN}(\omega)$ are presented in dotted and dashed lines. The dotted versions originate from the portion around ω_0 and the dashed versions from the portion around $-\omega_0$. From the figure it follows that the shifted versions remain undistorted as long as they do not overlap. By inspection it is concluded that this is the case if $(k-1)\omega_s \leq 2\omega_L$ and $k\omega_s \geq 2\omega_H$, where ω_L and ω_H are respectively the lower and higher frequency bounds of $S_{NN}(\omega)$. Combining the two conditions yields

$$\frac{2\omega_H}{k} \leq \omega_s \leq \frac{2\omega_L}{k-1} \tag{5.72}$$

In order to find a minimum amount of data to be processed, one should look for the lowest value of ω_s that satisfies these conditions. The lowest value of ω_s corresponds to the largest possible value of k. In doing so one has to realize that

$$2 \leq k \leq \frac{\omega_H}{W} \quad \text{and} \quad W \leq \omega_L \tag{5.73}$$

The lower bound on k is induced by the upper bound of ω_s and the upper bound is set to prevent the sampling rate becoming higher than would follow from the baseband sampling theorem. For the same reason the condition on W was introduced in Equations (5.73).

Example 5.2:

Suppose that we want to sample a bandpass process of bandwidth 50 MHz and with a central frequency of 1 GHz. This means that $\omega_L/(2\pi) = 975$ MHz and $\omega_H/(2\pi) = 1025$ MHz. From

Equations (5.73) it can easily be seen that the maximum allowable value is $k = 20$. The limits of the sample frequency are $102.5\text{MHz} < \omega_s/(2\pi) < 102.63$ MHz, which follows from Equation (5.72).

The possible minimum values of the sample frequency are quite near to what is found when applying the conversion to the baseband method from Section 5.5.1, which results in a sampling rate of 100 MHz.

□

5.6 SUMMARY

The chapter starts with a summary of the description of deterministic bandpass signals, where the so-called quadrature components are introduced. Bandpass signals or processes mostly result from the modulation of signals. Several modulation methods are briefly and conceptually described. Different complex description methods are introduced, such as the analytical signal and the complex envelope. Moreover, the baseband equivalent transfer function and impulse response of passband systems are defined. When applying all these concepts, their relation to the physical meaning should always be kept in mind when interpreting results.

For bandpass processes all these concepts are further exploited. Both in time and frequency domains a number of properties of the quadrature components are presented, together with their relation to the bandpass process.

The probability density function of the amplitude and phase are derived for bandpass filtered white Gaussian noise. These are of importance when dealing with modulated signals that are disturbed by noise.

The measurement of spectra is done using a so-called spectrum analyser. Since this equipment uses heterodyning and subsequently passband filtering, we use what we learned earlier in this chapter to describe its operation. The instrument is a powerful tool in the laboratory.

Finally, we describe the sampling of passband processes; two methods are given. It is proved that the minimum sampling rate of these processes equals twice the bandwidth.

5.7 PROBLEMS

5.1 It is well known that baseband systems provide distortionless transmission if the amplitude of the transfer characteristic is constant and the phase shift a linear function of frequency over the frequency band of the information signal [6, 13].

Let us now consider the conditions for distortionless transmission via a bandpass system. We call the transmission distortionless when on transmission the signal shape is preserved, but a delay in it may be allowed. It will be clear that once more the amplitude characteristic of the system has to be constant over the signal band. Furthermore, assume that the phase characteristic of the transfer function over the information band can be written as

$$\phi(\omega) = -j\tau_0\omega_0 - j\tau_g(\omega - \omega_0)$$

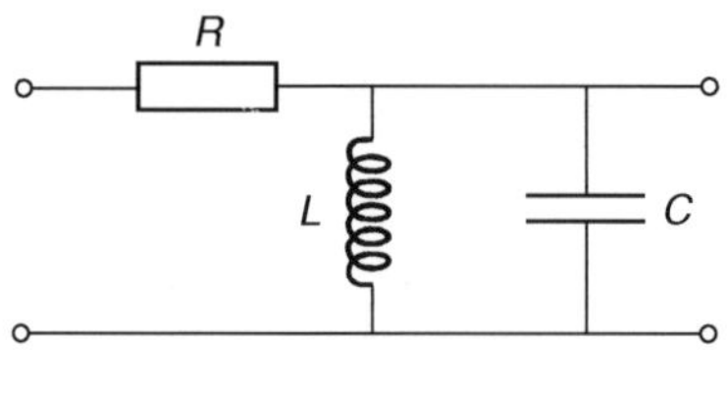

Figure 5.12

with τ_0 and τ_g constants.

(a) Show that, besides the amplitude condition, this phase condition provides distortionless transmission of signals modulated by the carrier frequency ω_0.

(b) What is the effect of τ_0 and τ_g on the output signal?

5.2 The circuit shown in Figure 5.12 is a bandpass filter.

(a) Derive the transfer function $H(\omega)$.

(b) Derive $|H(\omega)|$ and use Matlab to plot it for $LC = 0.01$ and $R\sqrt{C/L} = 10, 20, 50$.

(c) Let us call the resonance frequency ω_0. Determine the equivalent baseband transfer function and plot both its amplitude and phase for $R\sqrt{C/L} = 10, 50$, while preserving the value of LC.

(d) Determine for the values from (c) the phase delay (τ_0 from Problem 5.1) and plot the group delay (τ_g from Problem 5.1).

5.3 A bandpass process $N(t)$ has the following power spectrum:

$$S_{NN}(\omega) = \begin{cases} P\cos[\pi(\omega - \omega_0)/W], & -W/2 \leq \omega - \omega_0 \leq W/2 \\ P\cos[\pi(\omega + \omega_0)/W], & -W/2 \leq \omega + \omega_0 \leq W/2 \\ 0, & \text{all other values of } \omega \end{cases}$$

where P, W and $\omega_0 > W$ are positive, real constants.

(a) What is the power of $N(t)$?

(b) What is the power spectrum $S_{XX}(\omega)$ of $X(t)$ if $N(t)$ is represented as in Equation (5.20)?

(c) Calculate the cross-correlation function $R_{XY}(\tau)$.

(d) Are the quadrature processes $X(t)$ and $Y(t)$ orthogonal?

5.4 White noise with spectral density of $N_0/2$ is applied to an ideal bandpass filter with the central passband radial frequency ω_0 and bandwidth W.

(a) Calculate the autocorrelation function of the output process. Use Matlab to plot it.

(b) This output is sampled and the sampling instants are given as $t_n = n \times 2\pi/W$ with n integer values. What can be said about the sample values?

5.5 A bandpass process is represented as in Equation (5.20) and has the power spectrum according to Figure 5.13; assume that $\omega_1 > W$.

(a) Sketch $S_{XX}(\omega)$ and $S_{XY}(\omega)$ when $\omega_0 = \omega_1$.

(b) Repeat (a) when $\omega_0 = \omega_2$, where $\omega_1 - W/2 < \omega_2 < \omega_1$.

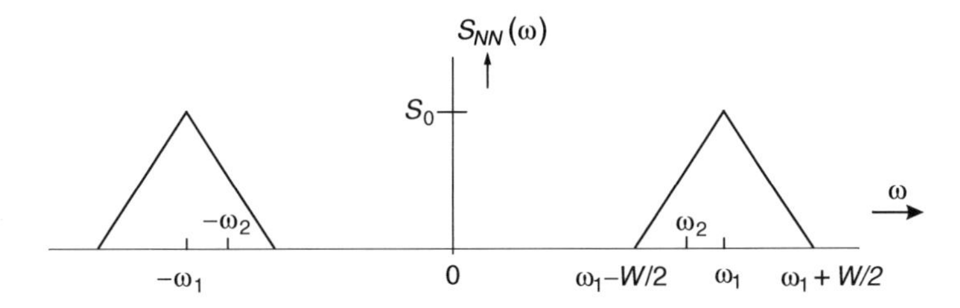

Figure 5.13

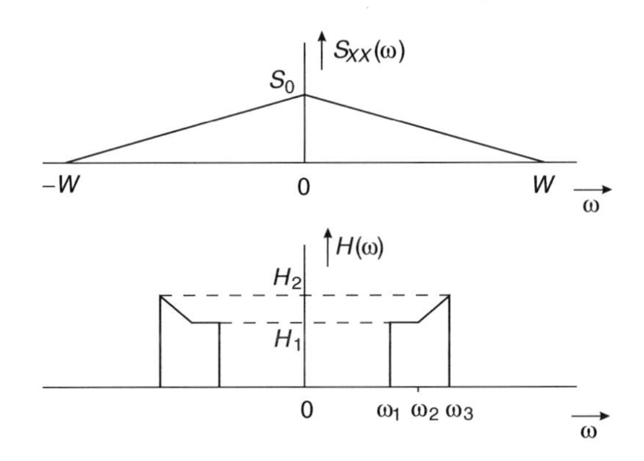

Figure 5.14

5.6 A wide-sense stationary process $X(t)$ has a power spectrum as depicted in the upper part of Figure 5.14. This process is applied to a filter with the transfer function $H(\omega)$ as given in the lower part of the figure. The data for the spectrum and filter are: $S_0 = 10^{-6}$, $W = 2\pi \times 10^7$, $\omega_1 = 2\pi \times 0.4 \times 10^7$, $\omega_2 = 2\pi \times 0.5 \times 10^7$, $\omega_3 = 2\pi \times 0.6 \times 10^7$, $H_1 = 2$ and $H_2 = 3$.

(a) Determine the power spectrum of the output.

(b) Use Matlab to plot the spectra of the quadrature components of the output when $\omega_0 = \omega_1$.

(c) Calculate the power of the output process.

5.7 A wide-sense stationary white Gaussian process has a spectral density of $N_0/2$. This process is applied to the input of the linear time-invariant filter. The filter has a bandpass characteristic with the transfer function

$$H(\omega) = \begin{cases} 1, & \omega_0 - W/2 < |\omega| < \omega_0 + W/2 \\ 0, & \text{elsewhere} \end{cases}$$

where $\omega_0 > W$.

(a) Sketch the transfer function $H(\omega)$.

(b) Calculate the mean value of the output process.

(c) Calculate the variance of the output process.

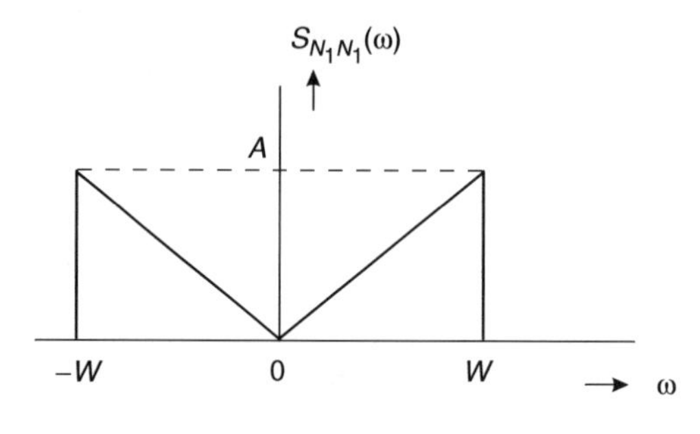

Figure 5.15

(d) Determine and dimension the probability density function of the output.

(e) Determine the power spectrum and the autocorrelation function of the output.

5.8 The wide-sense stationary bandpass noise process $N_1(t)$ has the central frequency ω_0. It is modulated by an harmonic carrier to form the process

$$N_2(t) = N_1(t)\cos(\omega_0 t - \Theta)$$

where Θ is independent of $N_1(t)$ and is uniformly distributed on the interval $(0, 2\pi]$.

(a) Show that $N_2(t)$ comprises both a baseband component and a bandpass component.

(b) Calculate the mean values and variances of these components, expressed in terms of the properties of $N_1(t)$.

5.9 The noise process $N_1(t)$ is wide-sense stationary. Its spectral density is given in Figure 5.15. By means of this process a new process $N_2(t)$ is produced according to

$$N_2(t) = N_1(t)\ \cos(\omega_0 t - \Theta) - N_1(t)\ \sin(\omega_0 t - \Theta)$$

where Θ is a random variable that is uniformly distributed on the interval $(0, 2\pi]$. Calculate and sketch the spectral density of $N_2(t)$.

5.10 Consider the stochastic process

$$N(t) = X(t)\cos(\omega_0 t - \Theta) - Y(t)\sin(\omega_0 t - \Phi)$$

with ω_0 a constant. The random variables Θ and Φ are independent of $X(t)$ and $Y(t)$ and uniformly distributed on the interval $(0, 2\pi]$. The spectra $S_{XX}(\omega)$, $S_{YY}(\omega)$ and $S_{XY}(\omega)$ are given in Figure 5.16, where $W_Y < W_X < \omega_0$ and in the right-hand picture the solid line is the real part of $S_{XY}(\omega)$ and the dashed line is its imaginary part.

(a) Determine and sketch the spectrum $S_{NN}(\omega)$ in the case where Θ and Φ are independent.

(b) Determine and sketch the spectrum $S_{NN}(\omega)$ in the case where $\Theta = \Phi$.

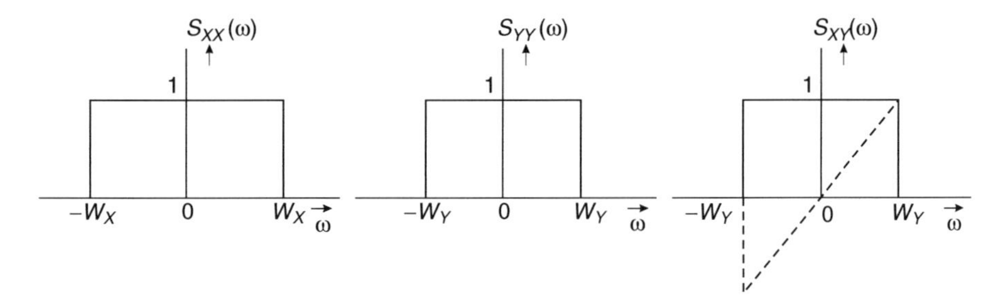

Figure 5.16

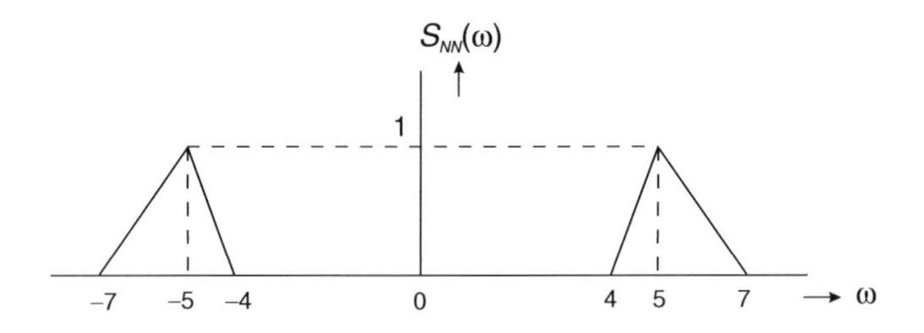

Figure 5.17

5.11 Consider the wide-sense stationary bandpass process

$$N(t) = X(t)\cos(\omega_0 t) - Y(t)\sin(\omega_0 t)$$

where $X(t)$ and $Y(t)$ are baseband processes. The spectra of these processes are

$$S_{XX}(\omega) = S_{YY}(\omega) = \begin{cases} 1, & |\omega| < W \\ 0, & |\omega| \geq W \end{cases}$$

and

$$S_{XY}(\omega) = \begin{cases} \mathrm{j}\frac{\omega}{W}, & |\omega| < W \\ 0, & |\omega| \geq W \end{cases}$$

where $W < \omega_0$.

(a) Sketch the spectra $S_{XX}(\omega)$, $S_{YY}(\omega)$ and $S_{XY}(\omega)$.

(b) Show how $S_{NN}(\omega)$ can be reconstructed from $S_{XX}(\omega)$ and $S_{XY}(\omega)$. Sketch $S_{NN}(\omega)$.

(c) Sketch the spectrum of the complex envelope of $N(t)$.

(d) Calculate the r.m.s. bandwidth of the complex envelope $Z(t)$.

5.12 A wide-sense stationary bandpass process has the spectrum as given in Figure 5.17. The characteristic frequency is $\omega_0 = 5$ rad/s.

(a) Sketch the power spectra of the quadrature processes.

(b) Are the quadrature processes uncorrelated?

(c) Are the quadrature processes independent?

5.13 A wide-sense stationary bandpass process is given by

$$N(t) = X(t)\cos(\omega_0 t) - Y(t)\sin(\omega_0 t)$$

where $X(t)$ and $Y(t)$ are independent random signals with an equal power of P_s and bandwidth $W < \omega_0$. These signals are received by a synchronous demodulator scheme as given in Figure 5.10; the lowpass filters are ideal filters, also of bandwidth W. The received signal is disturbed by additive white noise with spectral density $N_0/2$. Calculate the signal-to-noise ratios at the outputs.

5.14 The power spectrum of a narrowband wide-sense stationary bandpass process $N(t)$ needs to be measured. However, there is no spectrum analyser available that covers the frequency range of this process. Two product modulators are available, based on which the circuit of Figure 5.10 is constructed and the oscillator is tuned to the central frequency of $N(t)$. The LP filters allow frequencies smaller than W to pass unattenuated and block higher frequencies completely. By means of this set-up and a low-frequency spectrum analyser the spectra shown in Figure 5.18 are measured. Reconstruct the spectrum of $N(t)$.

5.15 The spectrum of a bandpass signal extends from 15 to 25 MHz. The signal is sampled with direct sampling.

(a) What is the range of possible sampling frequencies?

(b) How much higher is the minimum direct sampling frequency compared with the minimum frequency when conversion to baseband is applied.

(c) Compare the former two sampling frequencies with that following from the Nyquist baseband sampling theorem (Theorems 4 and 5).

5.16 A baseband signal of bandwidth 1 kHz is modulated on a carrier frequency of 8 kHz.

(a) Sketch the spectrum of the modulated bandpass signal.

(b) What is the minimum sampling frequency based on the Nyquist baseband sampling theorem (Theorems 4 and 5).

(c) What is the minimum sampling frequency based on direct sampling.

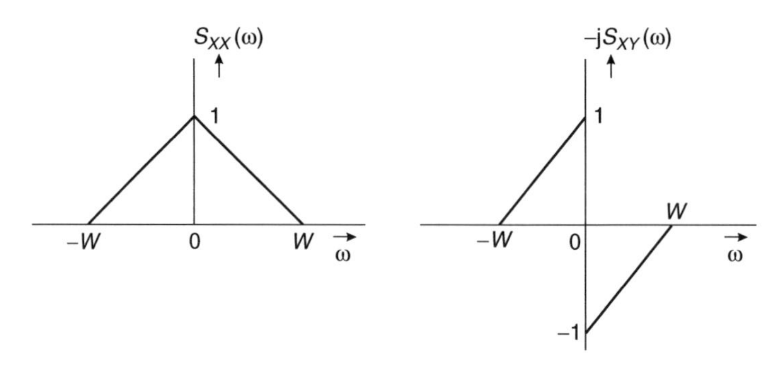

Figure 5.18

5.17 The transfer function of a discrete-time filter is given by

$$\tilde{H}(z) = \frac{1}{1 - 0.2z^{-1} + 0.95z^{-2}}$$

(a) Is this a stable system?

(b) Use Matlab's `freqz` to plot the absolute value of the transfer function.

(c) What type of filter is this? Search for the maximum value of the transfer function.

(d) Suppose that to the input of this filter a sinusoidal signal is applied with a frequency where the absolute value of the transfer function has its maximum value. Moreover, suppose that this signal is disturbed by wide-sense stationary white noise such that the signal-to-noise ratio amounts to 0 dB. Calculate the signal-to-noise ratio at the filter output.

(e) Explain the difference in signal-to-noise ratio improvement compared to that of Problem 4.28.

6

Noise in Networks and Systems

Many electrical circuits generate some kind of noise internally. The most well-known kind of noise is thermal noise produced by resistors. Besides this, several other kinds of noise sources can be identified, such as shot noise and partition noise in semiconductors. In this chapter we will describe the thermal noise generated by resistors, while shot noise is dealt with in Chapter 8. We shall show how internal noise sources can be transferred to the output terminals of a network, where the noise becomes observable to the outside world. For that purpose we shall consider the cascading of noisy circuits as well. In many practical situations, which we refer to in this chapter, a noise source can adequately be described on the basis of its power spectral density; this spectrum can be the result of a calculation or the result of a measurement as described in Section 5.4.

6.1 WHITE AND COLOURED NOISE

Realization of a wide-sense stationary noise process $N(t)$ is called white noise when the power spectral density of $N(t)$ has a constant value for all frequencies. Thus, it is a process for which

$$S_{NN}(\omega) = \frac{N_0}{2} \tag{6.1}$$

holds, with N_0 a real positive constant. By applying the inverse Fourier transform to this spectrum, the autocorrelation function of such a process is found to be

$$R_{NN}(\tau) = \frac{N_0}{2}\delta(\tau) \tag{6.2}$$

The name white noise was taken from optics, where white light comprises all frequencies (or equivalently all wavelengths) in the visible region.

Introduction to Random Signals and Noise W. van Etten

It is obvious that white noise cannot be a meaningful model for a noise source from a physical point of view. Looking at Equation (3.8) reveals that such a process would comprise an infinitely large amount of power, which is physically impossible. Despite the shortcomings of this model it is nevertheless often used in practice. The reason is that a number of important noise sources (see, for example, Section 6.2) have a flat spectrum over a very broad frequency range. Deviation from the white noise model is only observed at very high frequencies, which are of no practical importance.

The name coloured noise is used in situations where the power spectrum is not white. Examples of coloured noise spectra are lowpass, highpass and bandpass processes.

6.2 THERMAL NOISE IN RESISTORS

An important example of white noise is thermal noise. This noise is caused by thermal movement (or Brownian motion) of the free electrons in each electrical conductor. A resistor with resistance R at an absolute temperature of T has at its open terminals a noise voltage with a Gaussian probability density function with a mean value of zero and of which the power spectral density is

$$S_{VV}(\omega) = \frac{Rh|\omega|}{\pi\left[\exp\left(\frac{h|\omega|}{2\pi kT}\right) - 1\right]} \quad [\mathrm{V^2\,s}] \tag{6.3}$$

where

$$k = 1.38 \times 10^{-23} \quad [\mathrm{J/K}] \tag{6.4}$$

is the Boltzmann constant and

$$h = 6.63 \times 10^{-34} \quad [\mathrm{J\,s}] \tag{6.5}$$

is the Planck constant. Up until frequencies of 10^{12} Hz the expression (6.3) has an almost constant value, which gradually decreases to zero beyond that frequency. For useful frequencies in the radio, microwave and millimetre wavelength ranges, the power spectrum is white, i.e. flat. Using the well-known series expansion of the exponential in Equation (6.3), a very simple approximation of the thermal noise in a resistor is found.

Theorem 9

The spectrum of the thermal noise voltage across the open terminals of resistance R which is at the absolute temperature T is

$$S_{VV}(\omega) = 2kTR \quad [\mathrm{V^2\,s}] \tag{6.6}$$

This expression is much simpler than Equation (6.3) and can also be derived from physical considerations, which is beyond the scope of this text.

6.3 THERMAL NOISE IN PASSIVE NETWORKS

In Chapter 4 the response of a linear system to a stochastic process has been analysed. There it was assumed that the system itself was noise free, i.e. it does not produce noise itself. In the preceding section, however, we indicated that resistors produce noise; the same holds for semiconductor components such as transistors. Thus, if these components form part of a system, they will contribute to the noise at the output terminals. In this section we will analyse this problem. In doing so we will confine ourselves to the influence of thermal noise in passive networks. In a later section active circuits will be introduced.

As a model for a noisy resistor we introduce the equivalent circuit model represented in Figure 6.1. This model shows a noise-free resistance R in series with a noise process $V(t)$, for which Equation (6.6) describes the power spectral density. This scheme is called Thévenin's equivalent voltage model. From network theory we know that a resistor in series with a voltage source can also be represented as a resistance R in parallel with a current source. The magnitude of this current source is

$$I(t) = \frac{V(t)}{R} \tag{6.7}$$

In this way we arrive at the scheme given in Figure 6.2. This model is called Norton's equivalent current model. Using Equations (4.27) and (6.7) the spectrum of the current source is obtained.

> ***Theorem 10***
>
> The spectrum of the thermal noise current when short-circuiting a resistance R that is at the absolute temperature T is
>
> $$S_{II}(\omega) = \frac{2kT}{R} \quad [\mathrm{A}^2\,\mathrm{s}] \tag{6.8}$$

In both schemes of Figures 6.2 and 6.1, the resistors are assumed to be noise free.

When calculating the noise power spectral density at the output terminals of a network, the following method is used. Replace all noisy resistors by noise-free resistors in series with

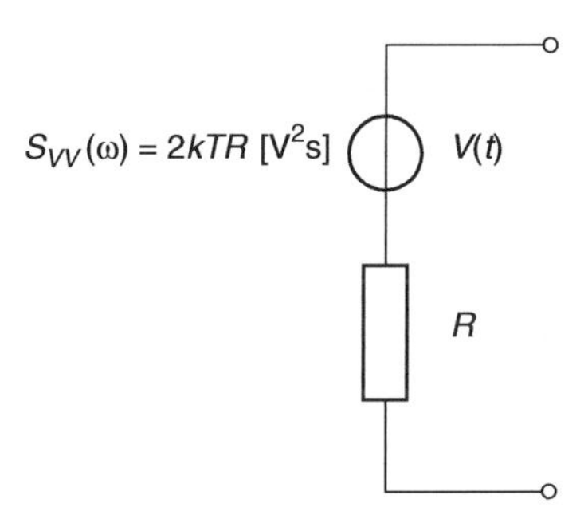

Figure 6.1 Thévenin equivalent voltage circuit model of a noisy resistor

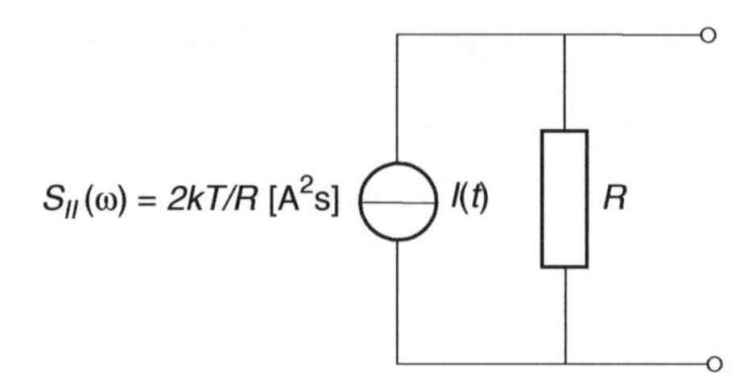

Figure 6.2 Norton equivalent current circuit model of a noisy resistor

a voltage source (according to Figure 6.1) or parallel with a current source (according to Figure 6.2). The schemes are equivalent, so it is possible to select the more convenient of the two schemes. Next, the transfer function from the voltage source or current source to the output terminals is calculated using network analysis methods. Invoking Equation (4.27), the noise power spectral density at the output terminals is found.

Example 6.1:

Consider the circuit presented in Figure 6.3. We wish to calculate the mean squared value of the voltage across the capacitor.

Express $V_c(\omega)$ in terms of V using the relationship

$$V_c(\omega) = H(\omega)\, V(\omega) \tag{6.9}$$

and

$$H(\omega) = \frac{V_c(\omega)}{V(\omega)} = \frac{\frac{1}{j\omega C}}{\frac{1}{j\omega C} + R} = \frac{1}{1 + j\omega RC} \tag{6.10}$$

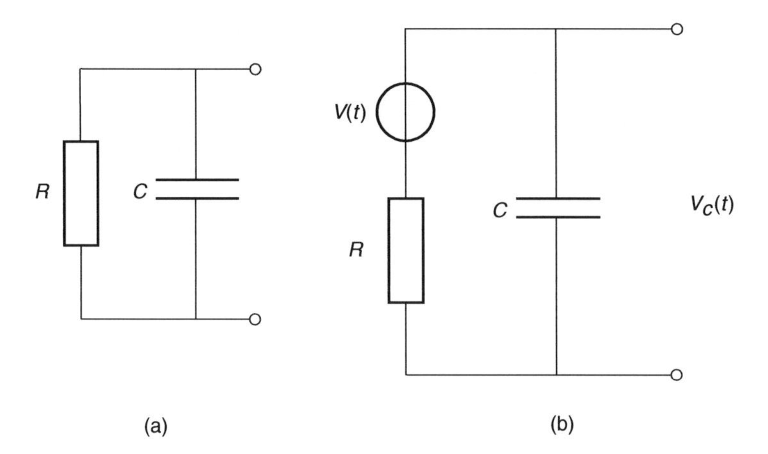

Figure 6.3 (a) Circuit to be analysed; (b) Thévenin equivalent model of the circuit

Invoking Equation (4.27), the power spectral density of $V_c(\omega)$ reads

$$S_{V_cV_c}(\omega) = 2kTR\frac{1}{1+\omega^2R^2C^2} \quad [\mathrm{V}^2\,\mathrm{s}] \tag{6.11}$$

and using Equation (4.28)

$$P_{V_c} = \frac{1}{2\pi}\int_{-\infty}^{\infty} \frac{2kTR}{1+\omega^2R^2C^2}\,\mathrm{d}\omega = \frac{kT}{C} \quad [\mathrm{V}^2\,\mathrm{s}] \tag{6.12}$$

□

When the network comprises several resistors, then these resistors will produce their noise independently from each other; namely the thermal noise is a consequence of the Brownian motion of the free electrons in the resistor material. As a rule the Brownian motion of electrons in one of the resistors will not be influenced by the Brownian motion of the electrons in different resistors. Therefore, at the output terminals the different spectra resulting from the several resistors in the circuit may be added.

Let us now consider the situation where a resistor is loaded by a second resistor (see Figure 6.4). If the loading resistance is called R_L, then similar to the method presented in Example 6.1, the power spectral density of the voltage V across R_L due to the thermal noise produced by R can be calculated. This spectral density is found by applying Equation (4.27) to the circuit of Figure 6.4, i.e. inserting the transfer function from the noise source to the load resistance

$$S_{VV}(\omega) = 2kTR\,|H(\omega)|^2 = 2kTR\frac{R_L^2}{(R+R_L)^2} \quad [\mathrm{V}^2\,\mathrm{s}] \tag{6.13}$$

Note the confusion that may arise here. When talking about the power of a stochastic process in terms of stochastic process theory, the expectation of the quadratic of the stochastic process is implied. This nomenclature is in accordance with Equation (6.13). However, when speaking about the physical concept of power, then conversion from the stochastic theoretical concept of power is required; this conversion will in general be simply multiplication by a constant factor. As for electrical power dissipated in a resistance R_L, we

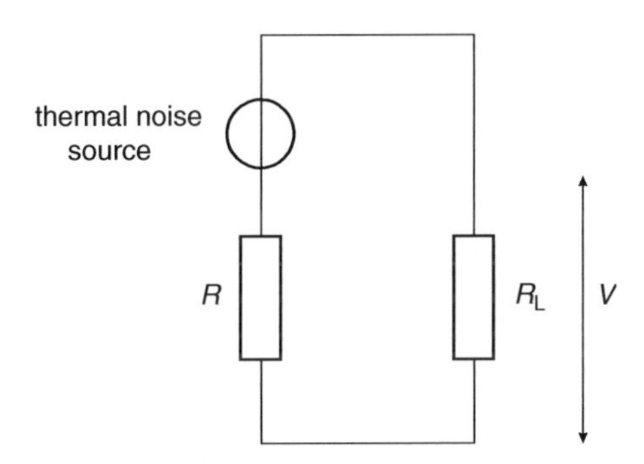

Figure 6.4 A resistance R producing thermal noise and loaded by a resistance R_L

have the formulas $P = \overline{V^2}/R_L = \overline{I^2}R_L$, and the conversion reads as

$$S_P(\omega) = \frac{1}{R_L} S_{VV}(\omega) = R_L\, S_{II}(\omega) \quad [\text{W s}] \tag{6.14}$$

For the spectral density of the electrical power that is dissipated in the resistor R_L we have

$$S_P(\omega) = 2kTR \frac{R_L}{(R + R_L)^2} \quad [\text{W s}] \tag{6.15}$$

It is easily verified that the spectral density given by Equation (6.15) achieves its maximum when $R = R_L$ and the density is

$$S_{P_{\max}}(\omega) \stackrel{\Delta}{=} S_a(\omega) = \frac{kT}{2} \quad [\text{W s}] \tag{6.16}$$

Therefore, the maximum power spectral density from a noisy resistor transferred to an external load amounts to $kT/2$, and this value is called the available spectral density. It can be seen that this spectral density is independent of the resistance value and only depends on temperature.

Analogously to Equation (6.6), white noise sources are in general characterized as

$$S_{VV}(\omega) = 2kT_e R_e \quad [\text{V}^2\,\text{s}] \tag{6.17}$$

In this representation the noise spectral density may have a larger value than the one given by Equation (6.6), due to the presence of still other noise sources than those caused by that particular resistor. We consider two different descriptions:

1. The spectral density is related to the value of the physical resistance R and we define $R_e = R$. In this case T_e is called the equivalent noise temperature; the equivalent noise temperature may differ from the physical temperature T.
2. The spectral density is related to the physical temperature T and we define $T_e = T$. In this case R_e is called the equivalent noise resistance; the equivalent noise resistance may differ from the physical value R of the resistance.

In networks comprising reactive components such as capacitors and coils, both the equivalent noise temperature and the equivalent noise resistance will generally depend on frequency. An example of this latter situation is elucidated when considering a generalization of Equation (6.6). For that purpose consider a circuit that only comprises passive components (R, L, C and an ideal transformer). The network may comprise several of each of these items, but it is assumed that all resistors are at the same temperature T. A pair of terminals constitute the output of the network and the question is: what is the power spectral density of the noise at the output of the circuit as a consequence of the thermal noise generated by the different resistors (hidden) in the circuit? The network is considered as a multiport; when the network comprises n resistors then we consider a multiport circuit with $n + 1$ terminal pairs. The output terminals are denoted by the terminal pair numbered 0.

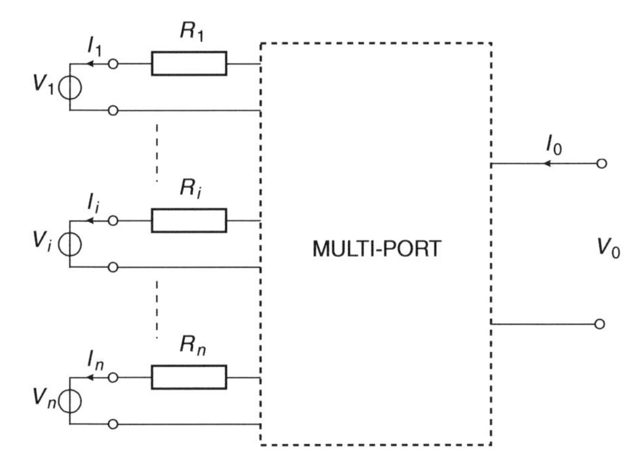

Figure 6.5 A network comprising n resistors considered as a multiport

Next, all resistors are put outside the multiport but connected to it by means of the terminal pairs numbered from 1 to n (see Figure 6.5). The relations between the voltages across the terminals and the currents flowing in or out of the multiport via the terminals are denoted using standard well-known network theoretical methods:

$$\begin{aligned} I_0 &= Y_{00}V_0 + \cdots + Y_{0i}V_i + \cdots + Y_{0n}V_n \\ \cdots &= \cdots \quad \cdots \quad \cdots \quad \cdots \quad \cdots \\ I_i &= Y_{i0}V_0 + \cdots + Y_{ii}V_i + \cdots + Y_{in}V_n \\ \cdots &= \cdots \quad \cdots \quad \cdots \quad \cdots \quad \cdots \\ I_n &= Y_{n0}V_0 + \cdots + Y_{ni}V_i + \cdots + Y_{nn}V_n \end{aligned} \tag{6.18}$$

Then it follows for the unloaded voltage at the output terminal pair 0 that

$$V_{0\text{open}} = \sum_{i=1}^{n} -\frac{Y_{0i}}{Y_{00}}V_i \tag{6.19}$$

The voltages, currents and admittances in Equations (6.18) and (6.19) are functions of ω. They represent voltages, currents and the relations between them when the excitation is a harmonic sine wave with angular frequency ω. Therefore, we may also write as an alternative to Equation (6.19)

$$V_{0\text{open}} = \sum_{i=1}^{n} H_i(\omega)V_i \tag{6.20}$$

When the voltage V_i is identified as the thermal noise voltage produced by resistor R_i, then the noise voltage at the output results from the superposition of all noise voltages originating from several resistors, each of them being filtered by a different transfer function $H_i(\omega)$. As observed before, we suppose the noise contribution from a certain resistor to be independent of these of all other resistors. Then the power spectral density of the output noise voltage is

$$S_{V_0V_0} = \sum_{i=1}^{n} |H_i(\omega)|^2 S_{V_iV_i} = \sum_{i=1}^{n} \left|\frac{Y_{0i}}{Y_{00}}\right|^2 S_{V_iV_i} = 2kT \sum_{i=1}^{n} \left|\frac{Y_{0i}}{Y_{00}}\right|^2 R_i \tag{6.21}$$

It appears that the summation may be substantially reduced. To this end consider the situation where the resistors are noise free (i.e. $V_i = 0$ for all $i \neq 0$) and where the voltage V_0 is applied to the terminal pair 0. From Equation (6.18) it follows in this case that

$$I_i = Y_{i0} V_0 \tag{6.22}$$

The dissipation in resistor R_i becomes

$$P_i = |I_i|^2 R_i = |Y_{i0}|^2 |V_0|^2 R_i \quad [\mathrm{W}] \tag{6.23}$$

As the multiport itself does not comprise any resistors, the total dissipation in the resistors has to be produced by the source that is applied to terminals 0, or

$$|I_0|^2 \mathrm{Re}\{Z_0\} = |V_0|^2 \sum_{i=1}^{n} |Y_{i0}|^2 R_i \tag{6.24}$$

where $\mathrm{Re}\{Z_0\}$ is the real part of the impedance of the multiport observed at the output terminal pair. As a consequence of the latter equation

$$\sum_{i=1}^{n} \frac{|Y_{i0}|^2}{|Y_{00}|^2} R_i = \mathrm{Re}\{Z_0\} \tag{6.25}$$

Passive networks are reciprocal, so that $Y_{i0} = Y_{0i}$. Substituting this into Equation (6.25) and the result from Equation (6.21) yields the following theorem.

Theorem 11

If in a passive network comprising several resistors, capacitors, coils and ideal transformers all resistors are at the same temperature T, then the voltage noise spectral density at the open terminals of this network is

$$S_{VV}(\omega) = 2kT\,\mathrm{Re}\{Z_0\} \quad [\mathrm{V}^2\,\mathrm{s}] \tag{6.26}$$

where Z_0 is the impedance of the network at the open terminal pair.

This generalization of Equation (6.6) is called Nyquist's theorem. Comparing Equation (6.26) with Equation (6.17) and if we take $T_e = T$, then the equivalent noise resistance becomes equal to $\mathrm{Re}\{Z_0\}$. When defining this quantity we emphasized that it can be frequency dependent. This is further elucidated when studying Example 6.1, which is presented in Figure 6.3.

Example 6.2:

Let us reconsider the problem presented in Example 6.1. The impedance at the terminals of Figure 6.3(a) reads

$$Z_0 = \frac{R}{1 + j\omega RC} \tag{6.27}$$

with its real part

$$\mathrm{Re}\{Z_0\} = \frac{R}{1 + \omega^2 R^2 C^2} \tag{6.28}$$

Substituting this expression into Equation (6.26) produces the voltage spectral density at the terminals

$$S_{V_c V_c}(\omega) = 2kT \frac{R}{1 + \omega^2 R^2 C^2} \quad [\mathrm{V}^2\,\mathrm{s}] \tag{6.29}$$

As expected, this is equal to the expression of Equation (6.11).

□

Equation (6.26) is a description according to Thévenin's equivalent circuit model (see Figure 6.1). A description in terms of Norton's equivalent circuit model is possible as well (see Figure 6.2). Then

$$S_{I_0 I_0}(\omega) = \frac{S_{V_0 V_0}}{|Z_0|^2} = 2kT\mathrm{Re}\{Y_0\} \quad [\mathrm{A}^2\,\mathrm{s}] \tag{6.30}$$

where

$$Y_0 \triangleq \frac{1}{Z_0} \tag{6.31}$$

Equation (6.30) presents the spectrum of the current that will flow through the shortcut that is applied to a certain terminal pair of a network. Here Y_0 is the admittance of the network at the shortcut terminal pair.

6.4 SYSTEM NOISE

The method presented in the preceding section can be applied to all noisy components in amplifiers and other subsystems that constitute a system. However, this leads to very extensive and complicated calculations and therefore is of limited value. Moreover, when buying a system the required details for such an analysis are not available as a rule. There is therefore a need for an alternative more generic noise description for (sub)systems in terms of relations between the input and output. Based on this, the quality of components, such as amplifiers, can be characterized in terms of their own noise contribution. In this way the noise behaviour of a system can be calculated simply and quickly.

6.4.1 Noise in Amplifiers

In general, amplifiers will contribute considerably to noise in a system, owing to the presence of noisy passive and active components in it. In addition, the input signals of amplifiers will in many cases also be disturbed by noise. We will start our analysis by considering ideal, i.e. noise-free, amplifiers. The most general equivalent scheme is presented in Figure 6.6. For the sake of simplifying the equations we will assume that all impedances in the scheme are real. A generalization to include reactive components is found in reference [4]. The amplifier has an input impedance of R_i, an output impedance of R_o and a transfer function of $H(\omega)$. The source has an impedance of R_s and generates as open voltage a wide-sense stationary stochastic voltage process V_s with the spectral density $S_{ss}(\omega)$. This process may represent noise or an information signal, or a combination (addition) of these types of processes. The available spectral density of this source is $S_s(\omega) = S_{ss}(\omega)/(4R_s)$. This follows from Equation (6.15) where the two resistances are set at the same value R_s. Using Equation (4.27), the available spectral density at the output of the amplifier is found to be

$$S_o(\omega) = \frac{S_{oo}(\omega)}{4R_o} = \frac{|H(\omega)|^2 S_{ii}(\omega)}{4R_o} = \frac{|H(\omega)|^2}{4R_o}\left(\frac{R_i}{R_s + R_i}\right)^2 S_{ss}(\omega) \tag{6.32}$$

The available power gain of the amplifier is defined as the ratio of the available spectral densities of the sources from Figure 6.6:

$$G_a(\omega) \triangleq \frac{S_o(\omega)}{S_s(\omega)} = \frac{S_{oo}(\omega)}{S_{ss}(\omega)}\frac{R_s}{R_o} = |H(\omega)|^2\left(\frac{R_i}{R_s + R_i}\right)^2 \frac{R_s}{R_o} \tag{6.33}$$

In case the impedances at the input and output are matched to produce maximum power transfer (i.e. $R_i = R_s$ and $R_o = R_L$), the practically measured gain will be equal to the available gain.

Now we will assume that the input source generates white noise, either from a thermal noise source or not, with an equivalent noise temperature of T_s. Then $S_s(\omega) = kT_s/2$ and the available spectral density at the output of the amplifier, supposed to be noise free, may be written as

$$S_o(\omega) = G_a(\omega)S_s(\omega) = G_a(\omega)\frac{kT_s}{2} \tag{6.34}$$

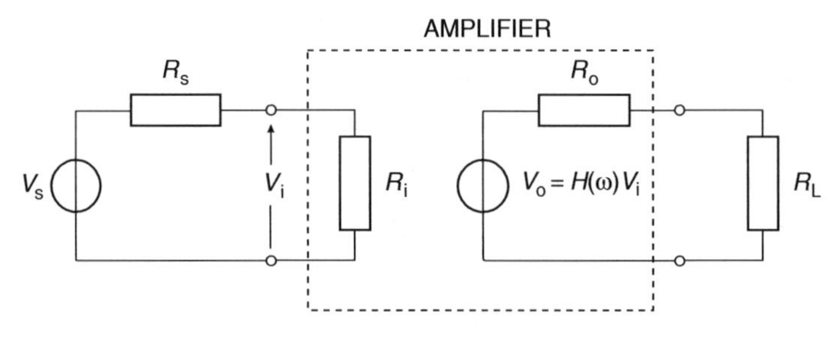

Figure 6.6 Model of an ideal (noise-free) amplifier with noise input

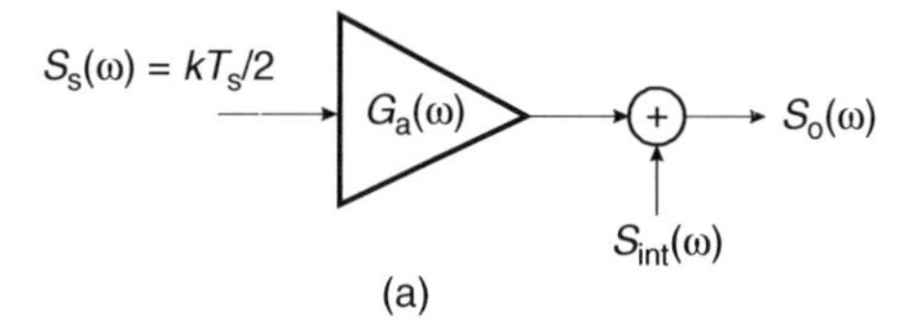

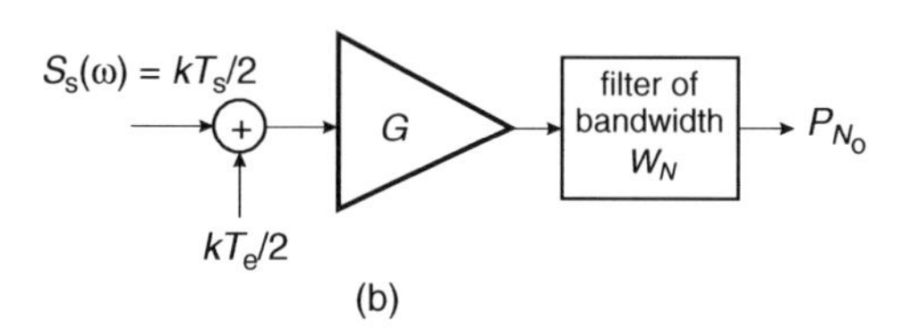

Figure 6.7 Block schematic of a noisy amplifier with the amplifier noise positioned (a) at the output or (b) at the input

From now on we will assume that the amplifier itself produces noise as well and it seems to be reasonable to suppose that the amplifier noise is independent of the noise generated by the source V_s. Therefore the available output spectral density is

$$S_o(\omega) = G_a(\omega)\frac{kT_s}{2} + S_{int}(\omega) \tag{6.35}$$

where $S_{int}(\omega)$ is the available spectral density at the output of the amplifier as a consequence of the noise produced by the internal noise sources present in the amplifier itself. The model that corresponds to this latter expression is drawn in Figure 6.7(a). The total available noise power at the output is found by integrating the output spectral density

$$P_{N_o} = \frac{1}{2\pi}\int_{-\infty}^{\infty} S_o(\omega)\,d\omega = \frac{1}{2\pi}\frac{kT_s}{2}\int_{-\infty}^{\infty} G_a(\omega)\,d\omega + \frac{1}{2\pi}\int_{-\infty}^{\infty} S_{int}(\omega)\,d\omega \tag{6.36}$$

This output noise power will be expressed in terms of the equivalent noise bandwidth (see Equation (4.51)) for the sake of simplifying the notation. For ω_0 we substitute that value for which the gain is maximal and we denote at that value $G_a(\omega_0) = G$. Then it is found that

$$\frac{1}{\pi}GW_N = \frac{1}{2\pi}\int_{-\infty}^{\infty} G_a(\omega)\,d\omega \tag{6.37}$$

Using this latter equation the first term of the right-hand side of Equation (6.36) can be written as $GkT_sW_N/(2\pi)$. In order to be able to write the second term of that equation in a similar way the effective noise temperature of the amplifier is defined as

$$T_e \triangleq \frac{1}{GkW_N}\int_{-\infty}^{\infty} S_{int}(\omega)\,d\omega \tag{6.38}$$

Based on this latter equation the total noise power at the output is written as

$$P_{N_o} = GkT_s\frac{W_N}{2\pi} + GkT_e\frac{W_N}{2\pi} = Gk(T_s + T_e)\frac{W_N}{2\pi} \tag{6.39}$$

It is emphasized that $W_N/(2\pi)$ represents the equivalent noise bandwidth in hertz. By the representation of Equation (6.39) the amplifier noise is in the model transferred to the input (see Figure 6.7(b)). In this way it can immediately be compared with the noise generated by the source at the input, which is represented by the first term in Equation (6.39).

6.4.2 The Noise Figure

Let us now consider a noisy device, an amplifier or a passive device, and let us suppose that the device is driven by a source that is noisy as well. The noise figure F of the device is defined as the ratio of the total available output noise spectral density (due to both the source and device) and the contribution to that from the source alone, in the later case supposing that the device is noise free. In general, the two noise contributions can have frequency-dependent spectral densities and thus the noise figure can also be frequency dependent. In that case it is called the spot noise figure. Another definition of the noise figure can be based on the ratio of the two total noise powers. In that case the corresponding noise figure is called the average noise figure. In many situations, however, the noise sources can be modelled as white sources. Then, based on the definition and Equation (6.39), it is found that

$$F = \frac{T_s + T_e}{T_s} = 1 + \frac{T_e}{T_s} \tag{6.40}$$

It will be clear that different devices can have different effective noise temperatures; this depends on the noise produced by the device. However, suppliers want to specify the quality of their devices for a standard situation. Therefore the standard noise figure for the situation where the source is at room temperature is defined as

$$F_s = 1 + \frac{T_e}{T_0}, \quad \text{with } T_0 = 290\,\text{K} \tag{6.41}$$

Thus for a very noisy device the effective noise temperature is much higher than room temperature ($T_e \gg T_0$) and $F_s \gg 1$. This does not mean that the physical temperature of the device is very high; this can and will, in general, be room temperature as well.

Especially for amplifiers, the noise figure can also be expressed in terms of signal-to-noise ratios. For that purpose the available signal power of the source is denoted by P_s, so that the signal-to-noise ratio at the input reads

$$\left(\frac{S}{N}\right)_s = \frac{P_s}{kT_s \frac{W_N}{2\pi}} \tag{6.42}$$

Note that the input noise power has only been integrated over the equivalent noise bandwidth W_N of the amplifier, although the input noise power is actually unlimited. This procedure is followed in order to be able to compare the input and output noise power based on the same bandwidth; for the output noise power it does not make any difference. It is an obvious choice to take for this bandwidth, the equivalent noise bandwidth, as this will reflect the actual noise power at the output. Furthermore, it is assumed that the signal spectrum is limited to the same bandwidth, so that the available signal power at the output is denoted as

$P_{so} = GP_s$. Using Equation (6.39) we find that the signal-to-noise ratio at the output is

$$\left(\frac{S}{N}\right)_o = \frac{GP_s}{P_{N_o}} = \frac{GP_s}{Gk(T_s + T_e)\frac{W_N}{2\pi}} \tag{6.43}$$

This signal-to-noise ratio is related to the signal-to-noise ratio at the input as

$$\left(\frac{S}{N}\right)_o = \frac{P_s}{(1+\frac{T_e}{T_s})kT_s\frac{W_N}{2\pi}} = \frac{1}{1+\frac{T_e}{T_s}}\left(\frac{S}{N}\right)_s \tag{6.44}$$

As the first factor in this expression is always smaller than 1, the signal-to-noise ratio is always deteriorated by the amplifier, which may not surprise us. This deterioration depends on the value of the effective noise temperature compared to the equivalent noise temperature of the source. If, for example, $T_e \ll T_s$, then the signal-to-noise ratio will hardly be reduced and the amplifier behaves virtually as a noise-free component. From Equation (6.44), it follows that

$$\frac{\left(\frac{S}{N}\right)_s}{\left(\frac{S}{N}\right)_o} = 1 + \frac{T_e}{T_s} = F \tag{6.45}$$

Note that the standard noise figure is defined for a situation where the source is at room temperature. This should be kept in mind when determining F by means of a measurement. Suppliers of amplifiers provide the standard noise figure as a rule in their data sheets, mostly presented in decibels (dB).

Sometimes, the noise figure is defined as the ratio of the two signal-to-noise ratios given in Equation (6.45). This can be done for amplifiers but can cause problems when considering a cascade of passive devices such as attenuators, since in that case input and output are not isolated and the load impedance of the source device is also determined by the load impedance of the devices.

Example 6.3:

As an interesting and important example, we investigate the noise figure of a passive two-port device such as a cable or an attenuator. Since the two-port device is passive it is reasonable to suppose that the power gain is smaller than 1 and denoted as $G = 1/L$, where L is the power loss of the two-port device. The signal-to-noise ratio at the input is written as in Equation (6.42), while the output signal power is by definition $P_{so} = P_s/L$. The passive two-port device is assumed to be at temperature T_a. The available spectral density of the output noise due to the noise contribution of the two-port device itself is determined by the impedance of the output terminals, according to Theorem 11. This spectral density is $kT_a/2$. The contribution of the source to the output available spectral density is $kT_s/(2L)$. However, since the resistance R_s of the input circuit is part of the impedance that is observed at the output terminals, the portion $kT_a/(2L)$ of its noise contribution to the output is already involved in the noise $kT_a/2$, which follows from the theorem. Only compensation for the difference in temperature is needed; i.e. we have to include an extra portion $k(T_s - T_a)/(2L)$.

Now the output signal-to-noise ratio becomes

$$\left(\frac{S}{N}\right)_{\mathrm{o}} = \frac{\frac{P_{\mathrm{s}}}{L}}{\left[kT_{\mathrm{a}} + \frac{k}{L}(T_{\mathrm{s}} - T_{\mathrm{a}})\right]\frac{W_N}{2\pi}} \tag{6.46}$$

After some simple calculations the noise figure follows from the definition

$$F_{\mathrm{s}} = 1 + (L-1)\frac{T_{\mathrm{a}}}{T_{\mathrm{s}}}, \quad \text{with} \quad T_{\mathrm{s}} = T_0 \tag{6.47}$$

When the two-port device is at room temperature this expression reduces to

$$F_{\mathrm{s}} = L \tag{6.48}$$

It is therefore concluded that the noise figure of a passive two-port device equals its power loss.

□

6.4.3 Noise in Cascaded Systems

In this subsection we consider the cascade connection of systems that may comprise several noisy amplifiers and other noisy components. We look for the noise properties of such a cascade connection, expressed as the parameters of the individual components as they are developed in the preceding subsection. In order to guarantee that the maximum power transfer occurs from one device to another, we assume that the impedances are matched; i.e. the input impedance of a device is the complex conjugate (see Problem 6.7) of the output impedance of the driving device. For the time being and for the sake of better understanding we only consider here a simple configuration consisting of the cascade of two systems (see Figure 6.8). The generalization to a cascade of more than two systems is quite simple, as will be shown later on. In the figure the relevant quantities of the two systems are indicated; they are the maximum power gain G_i, the effective noise temperature T_{ei} and the equivalent noise bandwidth W_i. The subscripts i refer to system 1 for $i = 1$ and to system 2 for $i = 2$, while the noise first enters system 1 and the output of system 1 is connected to the input of system 2 (see Figure 6.8). We assume that the passband of system 2 is completely encompassed by that of system 1, and as a consequence $W_2 \leq W_1$. This condition guarantees that all systems contribute to the output noise via the same bandwidth. Therefore, the equivalent noise bandwidth is equal to that of system 2:

$$W_N = W_2 \tag{6.49}$$

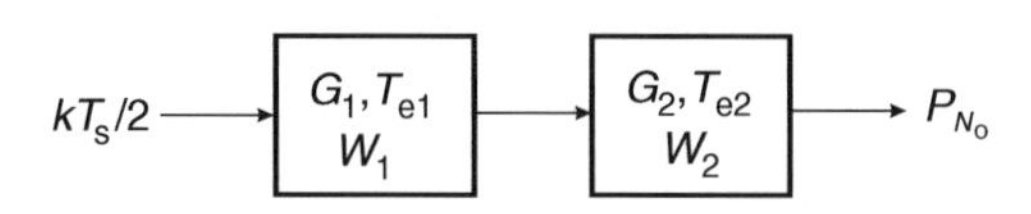

Figure 6.8 Cascade connection of two noisy two-port devices

The gain of the cascade is described by the product

$$G = G_1 G_2 \tag{6.50}$$

The total output noise consists of three contributions: the noise of the source that is amplified by both systems, the internal noise produced by system 1 and which is amplified by system 2 and the internal noise of system 2. Therefore, the output noise power is

$$P_{N_o} = (GkT_s + G_2G_1kT_{e1} + G_2kT_{e2})\frac{W_N}{2\pi} = Gk\left(T_s + T_{e1} + \frac{T_{e2}}{G_1}\right)\frac{W_N}{2\pi} \tag{6.51}$$

where the temperature expression

$$T_{sys} \triangleq T_s + T_{e1} + \frac{T_{e2}}{G_1} \tag{6.52}$$

is called the system noise temperature. From this it follows that the effective noise temperature of the cascade of the two-port devices in Figure 6.8 (see Equation (6.39)) is

$$T_e = T_{e1} + \frac{T_{e2}}{G_1} \tag{6.53}$$

and the noise figure of the cascade is found by inserting this equation into Equation (6.41), to yield

$$F_s = 1 + \frac{T_{e1}}{T_0} + \frac{T_{e2}}{G_1 T_0} = F_{s1} + \frac{F_{s2} - 1}{G_1} \tag{6.54}$$

Repeated application of the given method yields the effective noise temperature

$$T_e = T_{e1} + \frac{T_{e2}}{G_1} + \frac{T_{e3}}{G_1 G_2} + \cdots \tag{6.55}$$

and from that the noise figure of a cascade of three or more systems is

$$F_s = 1 + \frac{T_{e1}}{T_0} + \frac{T_{e2}}{G_1 T_0} + \frac{T_{e3}}{G_1 G_2 T_0} + \cdots = F_{s1} + \frac{F_{s2} - 1}{G_1} + \frac{F_{s3} - 1}{G_1 G_2} + \cdots \tag{6.56}$$

These two equations are known as the Friis formulas. From these formulas it is concluded that in a cascade connection the first stage plays a crucial role with respect to the noise behaviour; namely the noise from this first stage fully contributes to the output noise, whereas the noise from the next stages is to be reduced by a factor equal to the gain that precedes these stages. Therefore, in designing a system consisting of a cascade, the first stage needs special attention; this stage should show a noise figure that is as low as possible and a gain that is as large as possible. When the gain of the first stage is large, the effective noise temperature and noise figure of the cascade are virtually determined by those of the first stage. Following stages can provide further gain and eventual filtering, but will hardly influence the noise performance of the cascade. This means that the design demands of these stages can be relaxed.

Suppose that the first stage is a passive two-port device (e.g. a connection cable) with loss L_1. Inserting $G_1 = 1/L_1$ into Equation (6.56) yields

$$\begin{aligned} F_s &= L_1 + L_1(F_{s2} - 1) + L_1 \frac{F_{s3} - 1}{G_2} + \cdots \\ &= L_1 F_{s2} + L_1 \frac{F_{s3} - 1}{G_2} + \cdots \end{aligned} \tag{6.57}$$

Such a situation always causes the signal-to-noise ratio to deteriorate severely as the noise figure of the cascade consists mainly of that of the second (amplifier) stage multiplied by the loss of the passive first stage. When the second stage is a low-noise amplifier, this amplifier cannot repair the deterioration introduced by the passive two-port device of the first stage. Therefore, in case a lossy cable is needed to connect a low-noise device to processing equipment, the source first has to be amplified by a low-noise amplifier before applying it to the connection cable. This is elucidated by the next example.

Example 6.4:

Consider a satellite antenna that is connected to a receiver by means of a coaxial cable and an amplifier. The connection scheme and data of the different components are given in Figure 6.9. The antenna noise is determined by the low effective noise temperature of the dark sky (30 K) and produces an information signal power of −90 dBm in a bandwidth of 1 MHz at the input of the cable, which is at room temperature. All impedances are such that all the time maximum power transfer occurs. The receiver needs at least a signal-to-noise ratio of 17 dB. The question is whether the cascade can meet this requirement.

The signal power at the input of the receiver is

$$P_{sr} = (-90 - 2 + 60)\,\text{dBm} = -32\,\text{dBm} \Rightarrow 0.63\,\mu\text{W} \tag{6.58}$$

Using Equation (6.51), the noise power at the input of the receiver is

$$P_{N_0} = G_{\text{ampl}} G_{\text{coax}} k\, T_{\text{sys}} \frac{W_N}{2\pi} \tag{6.59}$$

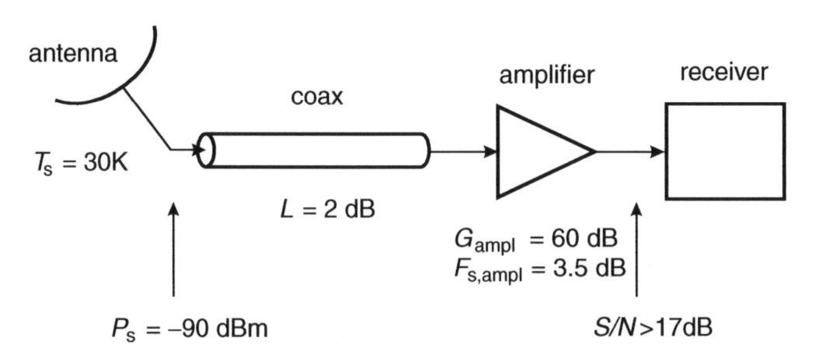

Figure 6.9 Satellite receiving circuit

On the linear scale, $G_{\text{ampl}} = 10^6$ and $G_{\text{coax}} = 0.63$. The effective noise temperatures of the coax and amplifier, according to Equation (6.41), are

$$\begin{aligned} T_{\text{e,coax}} &= T_0(F_{\text{s,coax}} - 1) = 290(1.58 - 1) = 168\,\text{K} \\ T_{\text{e,ampl}} &= T_0(F_{\text{s,ampl}} - 1) = 290(2.24 - 1) = 360\,\text{K} \end{aligned} \tag{6.60}$$

Inserting the numerical data into Equation (6.59) yields

$$P_{N_0} = 10^6 \times 0.63 \times 1.38 \times 10^{-23}\left(30 + 168 + \frac{360}{0.63}\right) \times 10^6 = 6.7 \times 10^{-9} \tag{6.61}$$

The ratio of P_{sr} and P_{N_0} produces the signal-to-noise ratio

$$\frac{S}{N} = \frac{P_{\text{sr}}}{P_{N_0}} = \frac{0.63 \times 10^{-6}}{6.7 \times 10^{-9}} = 94 \Rightarrow 19.7\,\text{dB} \tag{6.62}$$

It is concluded that the cascade satisfies the requirement of a minimum signal-to-noise ratio of 17 dB.

□

Although the suppliers characterize the components by the noise figure, in calculations as given in this example it is often more convenient to work with the effective noise temperatures in the way shown. Equation (6.41) gives a simple relation between the two data.

From the example it is clear that the coaxial cable does indeed cause the noise of the amplifier to be dominant.

Example 6.5:

As a second example of the noise figure of cascaded systems, we consider two different optical amplifiers, namely the so-called Erbium-doped fibre amplifier (EDFA) and the semiconductor optical amplifier (SOA). The first type is actually a fibre and so the insertion in a fibre link will give small coupling losses, let us say 0.5 dB. The second type, being a semiconductor device, has smaller waveguide dimensions than that of a fibre, which causes relatively high loss, let us say a 3 dB coupling loss. From physical reasoning it follows that optical amplifiers have a minimum noise figure of 3 dB. Let us compare the noise figure when either amplifier is inserted in a fibre link, where each of them has an amplification of 30 dB. On insertion we can distinguish three stages: (1) the coupling from the transmission fibre to the amplifier, (2) the amplifier device itself (EDFA or SOA) and (3) the output coupling from the amplifier device to the fibre. Using Equation (6.57) and the given data of these stages, the noise figure and other relevant data of the insertion are summarized in Table 6.1; note that in this table all data are in dB (see Appendix B). It follows from these data that the noise figure on insertion of the SOA is approximately 2.5 dB worse than that of the EDFA. This is almost completely attributed to the higher coupling loss at the front end of the amplifier. The output coupling hardly influences this number; it only contributes to a lower net gain.

□

Table 6.1 Comparing different optical amplifiers

	EDFA	SOA	Unit
Gain of device	30	30	dB
Coupling loss (×2)	0.5	3	dB
Noise figure device	3	3	dB
Noise figure on insertion	3.5	6	dB
Net insertion gain	29	24	dB

These two examples clearly show that in a cascade it is of the utmost importance that the first component (the front end) consists of a low-noise amplifier with a high gain, so that the front end contributes little noise and reduces the noise contribution of the other components in the cascade.

6.5 SUMMARY

A stochastic process is called 'white noise' if its power spectral density has a constant value for all frequencies. From a physical point of view this is impossible; namely this would imply an infinitely large amount of power. The concept in the first instance is therefore only of mathematical and theoretical value and may probably be used as a model in a limited but practically very wide frequency range. This holds specifically for thermal noise that is produced in resistors. In order to analyse thermal noise in networks and systems, we introduced the Thévenin and Norton equivalent circuit models. They consist of an ideal, that is noise-free, resistor in series with a voltage source or in parallel with a current source. Then, using network theoretical methods and the results from Chapter 4, the noise at the output of the network can easily be described. Several resistors in a network are considered as independent noise sources, where the superposition principle may be applied. Therefore, the total output power spectral density consists of the sum of the output spectra due to the individual resistors.

Calculating the noise behaviour of systems based on all the noisy components requires detailed data of the constituting components. This leads to lengthy calculations and frequently the detailed data are not available. A way out is offered by noise characterization of (sub)systems based on their output data. These output noise data are usually provided by component suppliers. Important data in this respect are the effective noise temperature and/ or the noise figure. On the basis of these parameters, the influence of the subsystems on the noise performance of a cascade can be calculated. From such an analysis it appears that the first stage of a cascade plays a crucial role. This stage should contribute as little as possible to the output noise (i.e. it must have a low effective noise temperature or, equivalently, a low-noise figure) and a high gain. The use of cables and attenuators as a first stage has to be avoided as they strongly deteriorate the signal-to-noise ratio. Such components should be preceded by low-noise amplifiers with a high gain.

6.6 PROBLEMS

6.1 Consider the thermal noise spectrum given by Equation (6.3).

(a) For which frequency range will this given spectrum have a value larger than $0.9 \times 2kTR$ at room temperature?

(b) Use Matlab to plot this power spectral density as a function of frequency, for $R = 1\ \Omega$ at room temperature.

(c) What is the significance of thermal noise in the optical domain, if it is realized that the optical domain as it is used for optical communication runs to a maximum wavelength of 1650 nm?

6.2 A resistance R_1 is at absolute temperature T_1 and a second resistance R_2 is at absolute temperature T_2.

(a) What is the equivalent noise temperature of the series connection of these two resistances?

(b) If $T_1 = T_2 = T$ what in that case is the value of T_e?

6.3 Answer the same questions as in Problem 6.2 but now for the parallel connection of the two resistances.

6.4 Consider once more the circuit of Problem 6.2. A capacitor with capacitance C_1 is connected parallel to R_1 and a capacitor with capacitance C_2 is connected parallel to R_2.

(a) Calculate the equivalent noise temperature.

(b) Is it possible to select the capacitances such that T_e becomes independent of frequency?

6.5 A resistor with a resistance value of R is at temperature T kelvin. A coil is connected parallel to this resistor with a self-inductance L henry. Calculate the mean value of the energy that is stored in the coil as a consequence of thermal noise produced by the resistor.

6.6 An electrical circuit consists of a loop of three elements in series, two resistors and a capacitor. The capacitance is C farad and the resistances are R_1 and R_2 respectively. Resistance R_1 is at temperature T_1 K and resistance R_2 is at T_2 K. Calculate the mean energy stored in the capacitor as a consequence of the thermal noise produced by the resistors.

6.7 A thermal noise source has an internal impedance of $Z(\omega)$. The noise source is loaded by the load impedance $Z_l(\omega)$.

(a) Show that a maximum power transfer from the noise source to the load occurs if $Z_l = Z^*(\omega)$.

(b) In that case what is the available power spectral density?

6.8 A resistor with resistance R_1 is at absolute temperature T_1. A second resistor with resistance R_2 is at absolute temperature T_2. The resistors R_1 and R_2 are connected in parallel.

(a) What is the spectral density of the net amount of power that is exchanged between the two resistors?

(b) Does the colder of the two resistors tend to further cool down due to this effect or heat up? In other words does the system strive for temperature equalization or does it strive to increase the temperature differences?

(c) What is the power exchange if the two temperatures are of equal value?

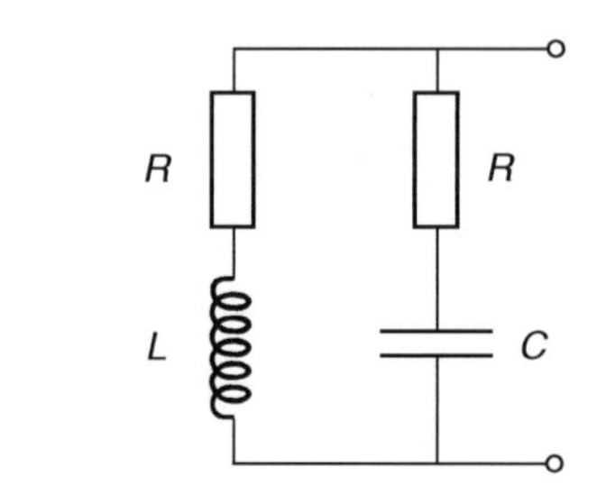

Figure 6.10

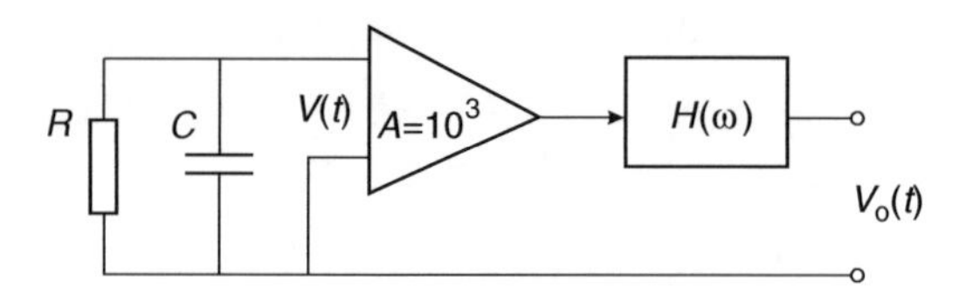

Figure 6.11

6.9 Consider the circuit in Figure 6.10, where all components are at the same temperature.

(a) The thermal noise produced by the resistors becomes manifest at the terminals. Suppose that the values of the components are such that the noise spectrum at the terminals is white. Derive the conditions in order for this to happen.

(b) What is in that case the impedance at the terminals?

6.10 Consider the circuit presented in Figure 6.11. The input impedance of the amplifier is infinitely high.

(a) Derive the expression for the spectral density $S_{VV}(\omega)$ of the input voltage $V(t)$ of the amplifier as a consequence of the thermal noise in the resistance R.

The lowpass filter $H(\omega)$ is ideal, i.e.

$$H(\omega) = \begin{cases} \exp(-j\omega\tau), & |\omega| \leq W \\ 0, & |\omega| > W \end{cases}$$

In the passband of $H(\omega)$ the constant $\tau = 1$ and the voltage amplification A can also be taken as constant and equal to 10^3. The amplifier does not produce any noise. The component values of the input circuit are $C = 200\,\text{nF}$ and $R = 1\,\text{k}\Omega$. The resistor is at room temperature so that $kT = 4 \times 10^{-21}$ W s.

(b) Calculate the r.m.s. value of the output voltage $V_o(t)$ of the filter in the case $W = 1/(RC)$.

6.11 Consider the circuit given in Figure 6.12. The data are as follows: $R = 50\ \Omega$, $L = 1\ \mu\text{H}$, $C = 400$ pF and $A = 100$.

(a) Calculate the spectral density of the noise voltage at the input of the amplifier as a consequence of the thermal noise produced by the resistors. Assume that these resistors are at room temperature and the other components are noise free.

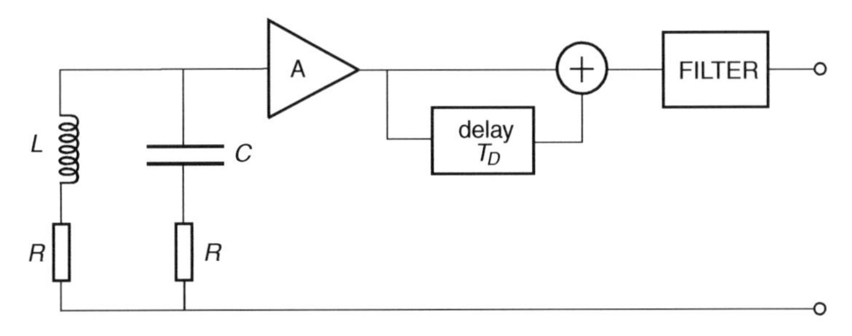

Figure 6.12

(b) Calculate the transfer function $H(\omega)$ from the output amplifier to the input filter.

(c) Calculate the spectral density of the noise voltage at the input of the filter.

(d) Calculate the r.m.s. value of the noise voltage at the filter output in the case where the filter is ideal lowpass with a transfer of 1 and a cut-off angular frequency $\omega_c = \pi/T_D$, where $T_D = 10\,\text{ns}$.

6.12 A signal source has a source impedance of 50 Ω and an equivalent noise temperature of 3000 K. This source is terminated by the input impedance of an amplifier, which is also 50 Ω. The voltage across this resistor is amplified and the amplifier itself is noise free. The voltage transfer function of the amplifier is

$$A(\omega) = \frac{100}{1 + j\omega\tau}$$

where $\tau = 10^{-8}$ s. The amplifier is at room temperature. Calculate the variance of the noise voltage at the output of the amplifier.

6.13 An amplifier is constituted from three stages with effective noise temperatures of $T_{e1} = 1300\,\text{K}$, $T_{e2} = 1750\,\text{K}$ and $T_{e3} = 2500\,\text{K}$, respectively, and where stage number 1 is the input stage, etc. The power gains amount to $G_1 = 20$, $G_2 = 10$ and $G_3 = 5$, respectively.

(a) Calculate the effective noise temperature of this cascade of amplifier stages.

(b) Explain why this temperature is considerably lower than T_{e2}, respectively T_{e3}.

6.14 An antenna has an impedance of 300 Ω. The antenna signal is amplified by an amplifier with an input impedance of 50 Ω. In order to match the antenna to the amplifier input impedance a resistor with a resistance of 300 Ω is connected in series with the antenna and parallel to the amplifier input a resistance of 50 Ω is connected.

(a) Sketch a block schematic of antenna, matching network and amplifier.

(b) Calculate the standard noise figure of the matching network.

(c) Do the resistances of 300 and 50 Ω provide matching of the antenna and amplifier? Support your answer by a calculation.

(d) Design a network that provides all the matching functionalities.

(e) What is the standard noise figure of this latter network? Compare this with the answer found for question (b).

6.15 An antenna is on top of a tall tower and is connected to a receiver at the foot of the tower by means of a cable. However, before applying the signal to the cable it is amplified. The amplifier has a power gain of 20 dB and a noise figure of $F = 3$ dB. The cable has a loss of 6 dB, while the noise figure of the receiver amounts to 13 dB. All impedances are matched; i.e. between components the maximum power transfer occurs.

(a) Calculate the noise figure of the system.

(b) Calculate the noise figure of the modified system where the amplifier is placed between the cable and the receiver at the foot of the tower instead of between the antenna and the cable at the top of the tower.

6.16 Reconsider Example 6.4. Interchange the order of the coaxial cable and the amplifier. Calculate the signal-to-noise ratio at the input of the receiver for this new situation.

6.17 An antenna is connected to a receiver via an amplifier and a cable. For proper operation the receiver needs at its input a signal-to-noise ratio of at least 20 dB. The amplifier is directly connected to the antenna and the cable connects the amplifier (power amplification of 60 dB) to the receiver. The cable has a loss of 1 dB and is at room temperature (290 K). The effective noise temperature of the antenna amounts to 50 K. The received signal is −90 dBm at the input of the amplifier and has a bandwidth of 10 MHz. All impedances are such that the maximum power transfer occurs.

(a) Present a block schematic of the total system and indicate in that sketch the relevant parameters.

(b) Calculate the signal power at the input of the receiver.

(c) The system designer can select one out of two suppliers for the amplifier. The suppliers A and B present the data given in Table 6.2. Which of the two amplifiers can be used in the system, i.e. on insertion of the amplifiers in the system which one will meet the requirement for the signal-to-noise ratio? Support your answer with a calculation.

6.18 Consider a source with a real source impedance of R_s. There are two passive networks as given in Figure 6.13. Resistance R_1 is at temperature T_1 K and resistance R_2 is at temperature T_2 K.

(a) Calculate the available power gain and standard noise factor when the circuit comprising R_1 is connected to the source.

(b) Calculate the available power gain and standard noise factor when the circuit comprising R_2 is connected to the source.

Table 6.2

	Supplier	
	A	B
Noise figure F (dB)	3.5	–
S/N reduction at the source temperature of 120 K (dB)	–	6

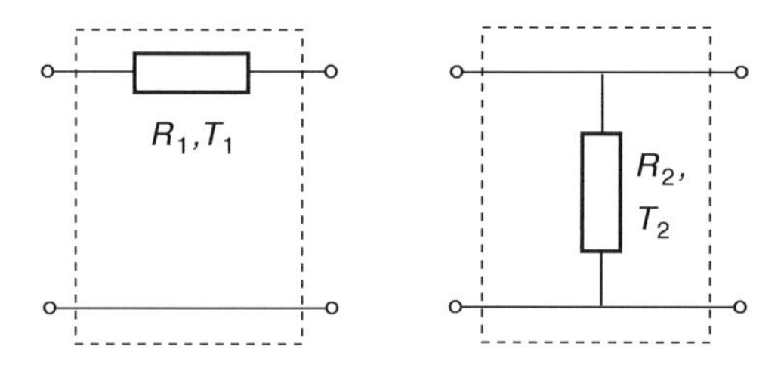

Figure 6.13

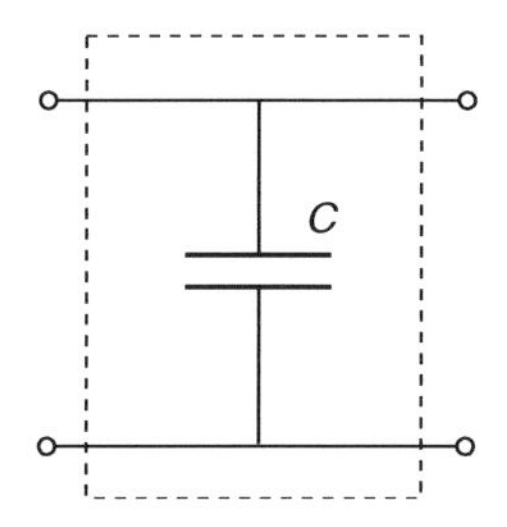

Figure 6.14

(c) Now assume that the two networks are cascaded where R_1 is connected to the source and R_2 to the output. Calculate the available gain and the standard noise figure of the cascade when connected to this source.

(d) Do the gains and the noise figures satisfy Equations (6.50) and (6.54), respectively? Explain your conclusion.

(e) Redo the calculations of the gain and noise figure when $R_s + R_1$ is taken as the source impedance for the second two-port device, i.e. the impedance of the source and the first two-port device as seen from the viewpoint of the second two-port device.

(f) Do the gains and the noise figures in case (e) satisfy Equations (6.50) and (6.54), respectively?

6.19 Consider a source with a complex source impedance of Z_s. This source is loaded by the passive network given in Figure 6.14.

(a) Calculate the available power gain and noise factor of the two-port device when it is connected to the source.

(b) Do the answers from (a) surprise you? If the answer is 'yes' explain why. If the answer is 'no' explain why not.

6.20 Consider the circuit given in Figure 6.15. This is a so-called 'constant resistance network'.

(a) Show that the input impedance of this circuit equals R_0 if $Z_1 Z_2 = R_0^2$ and the circuit is terminated by a resistance R_0.

(b) Calculate the available power gain and noise figure of the circuit (at temperature T kelvin) if the source impedance equals R_0 (which is at temperature T_s kelvin).

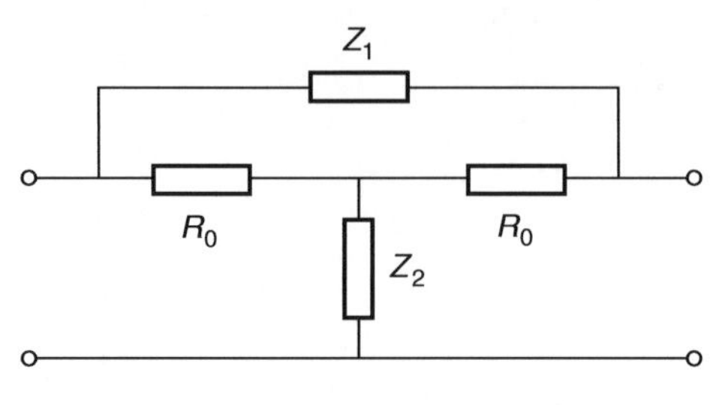

Figure 6.15

(c) Suppose that two of these circuits at different temperatures and with different gains are put in cascade and that the source impedance equals R_0 once more (also at temperature T_s kelvin). Calculate the overall available power gain and noise figure.

(d) Does the overall gain equal the product of the gains?

(e) Under what circumstances does the noise figure obey Equation (6.54)?

7
Detection and Optimal Filtering

Thus far the treatment has focused on the description of random signals and their analyses, and how these signals are transformed by linear time-invariant systems. In this chapter we take a somewhat different approach; namely starting with what is known about input processes and of system requirements we look for an optimum system. This means that we are going to perform system synthesis. The approach achieves an optimal reception of information signals that are corrupted by noise. In this case the input process consists of two parts, the information bearing or data signal and noise, and we may wonder what the optimal receiver or processing looks like, subject to some criterion.

When designing an optimal system three items play a crucial role. These are:

1. A description of the input noise process and the information bearing signal;
2. Conditions to be imposed on the system;
3. A criterion that defines optimality.

In the following we briefly comment on these items:

1. It is important to know the properties of the system inputs, e.g. the power spectral density of the input noise, whether it is wide-sense stationary, etc. What does the information signal look like? Are information signal and noise additive or not?
2. The conditions to be imposed on the system may influence performance of the receiver or the processing. We may require the system to be linear, time-invariant, realizable, etc. To start with and to simplify matters we will not bother about realizability. In specific cases it can easily be included.
3. The criterion will depend on the problem at hand. In the first instance we will consider two different criteria, namely the minimum probability of error in detecting data signals and the maximum signal-to-noise ratio. These criteria lead to an optimal linear filter called the matched filter. This name will become clear in the sequel. Although the criteria are quite different, we will show that there is a certain relationship in specific cases. In a third approach we will look for a filter that produces an optimum estimate of the

Introduction to Random Signals and Noise W. van Etten

realization of a stochastic process, which comes along with additive noise. In such a case we use the minimum mean-squared error criterion and end up with the so-called Wiener filter.

7.1 SIGNAL DETECTION

7.1.1 Binary Signals in Noise

Let us consider the transmission of a known deterministic signal that is disturbed by noise; the noise is assumed to be additive. This situation occurs in a digital communication system where, during successive intervals of duration T seconds, a pulse of known shape may arrive at the receiver (see the random data signal in Section 4.5). In such an interval the pulse has been sent or not. In accordance with Section 4.5 this transmitted random data signal is denoted by

$$Z(t) = \sum_n A_n\, p(t - nT) \tag{7.1}$$

Here A_n is randomly chosen from the set $\{0, 1\}$. The received signal is disturbed by additive noise. The presence of the pulse corresponds to the transmission of a binary digit '1' $(A_n = 1)$, whereas absence of the pulse in a specific interval represents the transmission of a binary digit '0' $(A_n = 0)$. The noise is assumed to be stationary and may originate from disturbance of the channel or has been produced in the front end of the receiver equipment. Every T seconds the receiver has to decide whether a binary '1' or a binary '0' has been sent. This decision process is called detection. Noise hampers detection and causes errors to occur in the detection process, i.e. '1's may be interpreted as '0's and vice versa.

During each bit interval there are two possible mutually exclusive situations, called hypotheses, with respect to the received signal $R(t)$:

$$\mathrm{H}_0: \quad R(t) = N(t), \qquad 0 \le t \le T \tag{7.2}$$

$$\mathrm{H}_1: \quad R(t) = p(t) + N(t), \qquad 0 \le t \le T \tag{7.3}$$

The hypothesis H_0 corresponds to the situation that a '0' has been sent $(A_n = 0)$. In this case the received signal consists only of the noise process $N(t)$. Hypothesis H_1 corresponds to the event that a '1' has been sent $(A_n = 1)$. Now the received signal comprises the known pulse shape $p(t)$ and the additive noise process $N(t)$. It is assumed that each bit occupies the $(0, T)$ interval. Our goal is to design the receiver such that in the detection process the probability of making wrong decisions is minimized. If the receiver decides in favour of hypothesis H_0 and it produces a '0', we denote the estimate of A_n by $\hat{A}_n$ and say that $\hat{A}_n = 0$. In case the receiver decides in favour of hypothesis H_1 and a '1' is produced, we denote $\hat{A}_n = 1$. Thus the detected bit $\hat{A}_n \in \{0, 1\}$. In the detection process two types of errors can be made. Firstly, the receiver decides in favour of hypothesis H_1, i.e. a '1' is detected $(\hat{A}_n = 1)$, whereas a '0' has been sent $(A_n = 0)$. The conditional probability of this event is $\mathrm{P}(\hat{A}_n = 1 \,|\, \mathrm{H}_0) = \mathrm{P}(\hat{A}_n = 1 \,|\, A_n = 0)$. Secondly, the receiver decides in favour of hypothesis H_0 $(\hat{A}_n = 0)$, whereas a '1' has been sent $(A_n = 1)$. The conditional probability of this event is $\mathrm{P}(\hat{A}_n = 0 \,|\, \mathrm{H}_1) = \mathrm{P}(\hat{A}_n = 0 \,|\, A_n = 1)$. In a long sequence of transmitted bits the prior probability of sending a '0' is given by P_0 and the prior probability of a '1' by P_1. We assume that these probabilities are known in the receiver. In accordance with the law of total probability the bit error probability is given by

$$\begin{aligned} P_e &= P_0\,P(\hat{A}_n = 1 \,|\, H_0) + P_1\,P(\hat{A}_n = 0 \,|\, H_1) \\ &= P_0\,P(\hat{A}_n = 1 \,|\, A_n = 0) + P_1\,P(\hat{A}_n = 0 \,|\, A_n = 1) \end{aligned} \tag{7.4}$$

This error probability is minimized if the receiver chooses the hypothesis with the highest conditional probability, given the process $R(t)$. It will be clear that the conditional probabilities of Equation (7.4) depend on the signal $p(t)$, the statistical properties of the noise $N(t)$ and the way the receiver processes the received signal $R(t)$. As far as the latter is concerned, we assume that the receiver converts the received signal $R(t)$ into K numbers (random variables), which are denoted by the K-dimensional random vector

$$\mathbf{r} = (r_1, r_2, \ldots, r_K) \tag{7.5}$$

The receiver chooses the hypothesis H_1 if $P(H_1 | \mathbf{r}) \geq P(H_0 | \mathbf{r})$, or equivalently $P_1\, f_{\mathbf{r}}(\mathbf{r} \,|\, H_1) \geq P_0\, f_{\mathbf{r}}(\mathbf{r} \,|\, H_0)$, since it follows from Bayes' theorem (reference [14]) that

$$P(H_i \,|\, \mathbf{r}) = \frac{P_i\, f_{\mathbf{r}}(\mathbf{r} \,|\, H_i)}{f_{\mathbf{r}}(\mathbf{r})}, \quad i = 0, 1 \tag{7.6}$$

From this it follows that the decision can be based on the so-called likelihood ratio

$$\Lambda(\mathbf{r}) \triangleq \frac{f_{\mathbf{r}}(\mathbf{r} \,|\, H_1)}{f_{\mathbf{r}}(\mathbf{r} \,|\, H_0)} \underset{H_0}{\overset{H_1}{\gtrless}} \Lambda_0 \triangleq \frac{P_0}{P_1} \tag{7.7}$$

In other words, hypothesis H_1 is chosen if $\Lambda(\mathbf{r}) > \Lambda_0$ and hypothesis H_0 is chosen if $\Lambda(\mathbf{r}) < \Lambda_0$. The quantity Λ_0 is called the decision threshold. In taking the decision the receiver partitions the vector space spanned by $\mathbf{r}$ into two parts, R_0 and R_1, called the decision regions. The boundary between these two regions is determined by Λ_0. In the region R_0 we have the relation $\Lambda(\mathbf{r}) < \Lambda_0$ and an observation of $\mathbf{r}$ in this region causes the receiver to decide that a binary '0' has been sent. An observation in the region R_1, i.e. $\Lambda(\mathbf{r}) \geq \Lambda_0$, makes the receiver decide that a binary '1' has been sent. The task of the receiver therefore is to transform the received signal $R(t)$ into the random vector $\mathbf{r}$ and determine to which of the regions R_0 or R_1 it belongs. Later we will go into more detail of this signal processing.

Example 7.1:

Consider the two conditional probability densities

$$f_r(r \,|\, H_0) = \frac{1}{\sqrt{2\pi}} \exp\left(-\frac{r^2}{2}\right) \tag{7.8}$$

$$f_r(r \,|\, H_1) = \tfrac{1}{2} \exp(-|r|) \tag{7.9}$$

and the prior probabilities

$$P_0 = P_1 = \tfrac{1}{2} \tag{7.10}$$

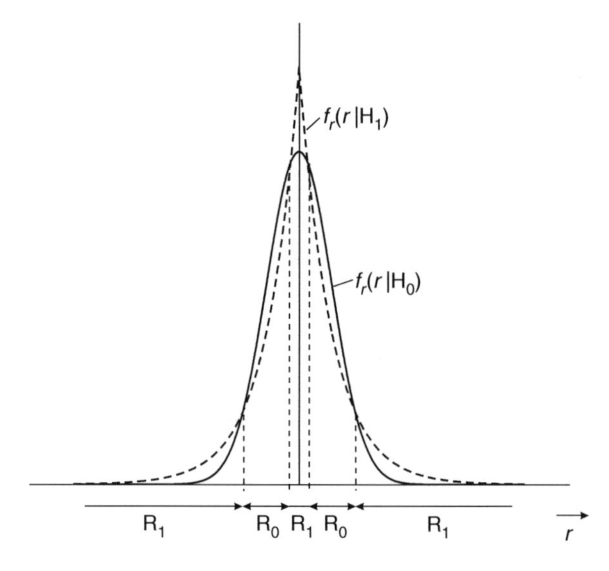

Figure 7.1 Conditional probability density functions of the example and the decision regions R_0 and R_1

Let us calculate the decision regions for this situation. By virtue of Equation (7.10) the decision threshold is set to one and the decision regions are found by equating the right-hand sides of Equations (7.8) and (7.9):

$$\frac{1}{\sqrt{2\pi}} \exp\left(-\frac{r^2}{2}\right) = \frac{1}{2}\exp(-|r|) \tag{7.11}$$

The two expressions are depicted in Figure 7.1. As seen from the figure, the functions are even symmetric and, confining to positive values of r, this equation can be rewritten as the quadratic

$$r^2 - 2r - 2\ln\frac{2}{\sqrt{2\pi}} = 0 \tag{7.12}$$

Solving this yields the roots $r_1 = 0.259$ and $r_2 = 1.741$. Considering negative r values produces the same negative values for the roots. Hence it may be concluded that the decision regions are described by

$$R_0 = \{r : 0.259 < |r| < 1.741\} \tag{7.13}$$

$$R_1 = \{r : (|r| < 0.259) \cup (|r| > 1.741)\} \tag{7.14}$$

□

One may wonder what to do when an observation is exactly at the boundaries of the decision regions. An arbitrary decision can be made, since the probability of this event approaches zero.

The conditional error probabilities in Equation (7.4) are written as

$$\mathrm{P}(\hat{A}_n = 1 \,|\, \mathrm{H}_0) = \mathrm{P}\{\Lambda(\mathbf{r}) \geq \Lambda_0 \,|\, \mathrm{H}_0\} = \int \underset{R_1}{\cdots} \int f_{\mathbf{r}}(\mathbf{r} \,|\, \mathrm{H}_0)\, \mathrm{d}r_1 \cdots \mathrm{d}r_K \tag{7.15}$$

$$\mathrm{P}(\hat{A}_n = 0 \,|\, \mathrm{H}_1) = \mathrm{P}\{\Lambda(\mathbf{r}) < \Lambda_0 \,|\, \mathrm{H}_1\} = \int \underset{R_0}{\cdots} \int f_{\mathbf{r}}(\mathbf{r} \,|\, \mathrm{H}_1)\, \mathrm{d}r_1 \cdots \mathrm{d}r_K \tag{7.16}$$

The minimum total bit error probability is found by inserting these quantities in Equation (7.4).

Example 7.2:

An example of a received data signal (see Equation (7.1)) has been depicted in Figure 7.2(a). Let us assume that the signal $R(t)$ is characterized by a single number r instead of a vector and that in the absence of noise ($N(t) \equiv 0$) this number is symbolically denoted by '0' (in the case of hypothesis H_0) or '1' (in the case of hypothesis H_1). Furthermore, assume that in the presence of noise $N(t)$ a stochastic Gaussian variable should be added to this characteristic number. For this situation the conditional probability density functions are given in Figure 7.2(a) upper right. In Figure 7.2(b) these functions are depicted once more, but now in a somewhat different way. The boundary that separates the decision regions R_0 and R_1 reduces to a single point. This point r_0 is determined by Λ_0. The bit error probability is now written as

$$\mathrm{P_e} = \mathrm{P}_0 \int_{r_0}^{\infty} f_r(r \,|\, \mathrm{H}_0)\, \mathrm{d}r + \mathrm{P}_1 \int_{-\infty}^{r_0} f_r(r \,|\, \mathrm{H}_1)\, \mathrm{d}r \tag{7.17}$$

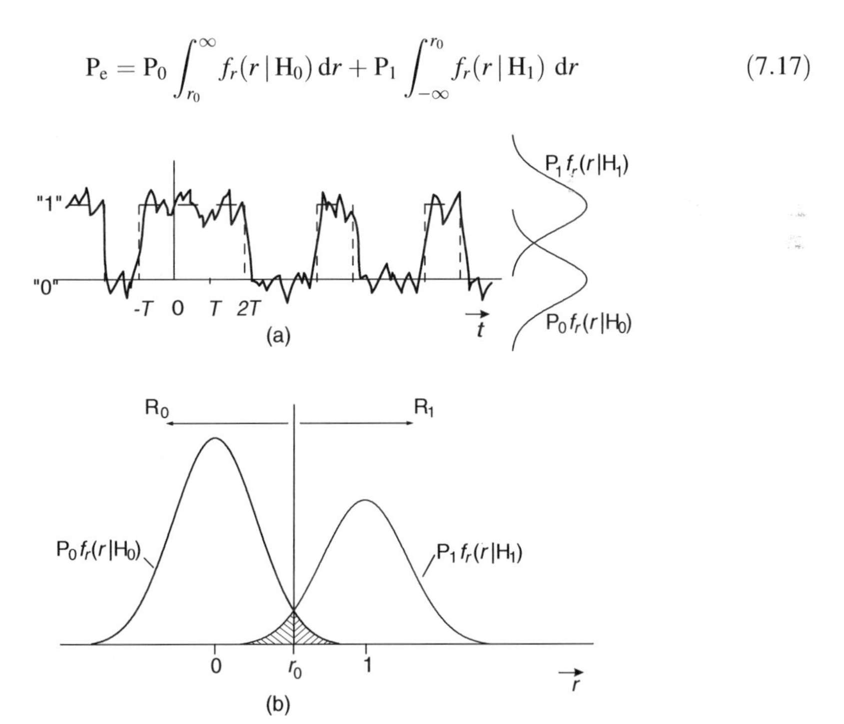

Figure 7.2 (a) A data signal disturbed by Gaussain noise and (b) the corresponding weighted (by the prior probabilities) conditional probability density functions and the decision regions R_0 and R_1

The first term on the right-hand side of this equation is represented by the right shaded region in Figure 7.2(b) and the second term by the left shaded region in this figure. The threshold value r_0 is to be determined such that P_e is minimized. To that end P_e is differentiated with respect to r_0

$$\frac{dP_e}{dr_0} = -P_0\, f_r(r_0 \,|\, H_0) + P_1\, f_r(r_0 \,|\, H_1) \tag{7.18}$$

When this expression is set equal to zero we once again arrive at Equation (7.7); in this way this equation has been deduced in an alternative manner. Now it appears that the optimum threshold value r_0 is found at the intersection point of the curves $P_0\, f_r(r \,|\, H_0)$ and $P_1\, f_r(r \,|\, H_1)$. If the probabilities P_0 and P_1 change but the probability density function of the noise $N(t)$ remains the same, then the optimum threshold value shifts in the direction of the binary level that corresponds to the shrinking prior probability.

Remembering that we considered the case of Gaussian noise, it is concluded that the integrals in Equation (7.17) can be expressed using the well-known Q function (see Appendix F), which is defined as

$$Q(x) \triangleq \frac{1}{\sqrt{2\pi}} \int_x^{\infty} \exp\left(-\frac{y^2}{2}\right) dy \tag{7.19}$$

This function is related to the erfc(·) function as follows:

$$Q(x) = \frac{1}{2}\operatorname{erfc}\left(\frac{x}{\sqrt{2}}\right) \tag{7.20}$$

Both functions are tabulated in many books or can be evaluated using software packages. They are presented graphically in Appendix F.

More details on the Gaussian noise case are presented in the next section.

□

7.1.2 Detection of Binary Signals in White Gaussian Noise

In this subsection we will assume that in the detection process as described in the foregoing the disturbing noise $N(t)$ has a Gaussian probability density function and a white spectrum with a spectral density of $N_0/2$. This latter assumption means that filtering has to be performed in the receiver. This is understood if we realize that a white spectrum implies an infinitely large noise variance, which leads to problems in the integrals that appear in Equation (7.17). Filtering limits the extent of the noise spectrum to a finite frequency band, thereby limiting the noise variance to finite values and thus making the integrals well defined.

The received signal $R(t)$ is processed in the receiver to produce the vector $\mathbf{r}$ in a signal space $\{\varphi_k(t)\}$ that completely describes the signal $p(t)$ and is assumed to be an orthonormal set (see Appendix A)

$$r_k = \int_0^T \phi_k(t)\, R(t)\, dt, \quad k = 1, \ldots, K \tag{7.21}$$

As the operation given by Equation (7.21) is a linear one, it can be applied to the two terms of Equation (7.3) separately, so that

$$r_k = p_k + n_k, \quad k = 1, \ldots, K \tag{7.22}$$

with

$$p_k \triangleq \int_0^T \phi_k(t)\, p(t)\, \mathrm{d}t, \quad k = 1, \ldots, K \tag{7.23}$$

and

$$n_k \triangleq \int_0^T \phi_k(t)\, N(t)\, \mathrm{d}t, \quad k = 1, \ldots, K \tag{7.24}$$

In fact, the processing in the receiver converts the received signal $R(t)$ into a vector $\mathbf{r}$ that consists of the sum of the deterministic signal vector $\mathbf{p}$, of which the elements are given by Equation (7.23), and the noise vector $\mathbf{n}$, of which the elements are given by Equation (7.24). As $N(t)$ has been assumed to be Gaussian, the random variables n_k will be Gaussian as well. This is due to the fact that when a linear operation is performed on a Gaussian variable the Gaussian character of the random variable is maintained. It follows from Appendix A that the elements of the noise vector are orthogonal and all of them have the same variance $N_0/2$. In fact, the noise vector $\mathbf{n}$ defines the relevant noise (Appendix A and reference [14]).

Considering the case of binary detection, the conditional probability density functions for the two hypotheses are now

$$f_{\mathbf{r}}(\mathbf{r} \mid \mathrm{H}_0) = \frac{1}{\sqrt{(\pi N_0)^K}} \exp\left(-\frac{1}{N_0}\sum_{k=1}^{K} r_k^2\right) \tag{7.25}$$

and

$$f_{\mathbf{r}}(\mathbf{r} \mid \mathrm{H}_1) = \frac{1}{\sqrt{(\pi N_0)^K}} \exp\left(-\frac{1}{N_0}\sum_{k=1}^{K} (r_k - p_k)^2\right) \tag{7.26}$$

Using Equation (7.7), the likelihood ratio is written as

$$\begin{aligned}\Lambda(\mathbf{r}) &= \exp\left[-\frac{1}{N_0}\sum_{k=1}^{K}(r_k - p_k)^2 + \frac{1}{N_0}\sum_{k=1}^{K} r_k^2\right] \\ &= \exp\left(\frac{2}{N_0}\sum_{k=1}^{K} p_k r_k - \frac{1}{N_0}\sum_{k=1}^{K} p_k^2\right)\end{aligned} \tag{7.27}$$

In Appendix A it is shown that the term $\sum p_k^2$ represents the energy E_p of the deterministic signal $p(t)$. This quantity is supposed to be known at the receiver, so that the only quantity that depends on the transmitted signal consists of the summation over $p_k r_k$. By means of signal processing on the received signal $R(t)$, the value of this latter summation should be determined. This result represents a sufficient statistic [3,9] for detecting the transmitted data in an optimal way. A statistic is an operation on an observation, which is presented by a function or functional. A statistic is said to be sufficient if it preserves all information that is relevant for estimating the data. In this case it means that in dealing with the noise component of **r** the irrelevant noise components may be ignored (see Appendix A or reference [14]). This is shown as follows:

$$\begin{aligned}\sum_k p_k r_k &= \sum_k p_k(p_k + n_k) \\ &= \int_0^T \sum_k \phi_k(t) p(t)(p_k + n_k)\,\mathrm{d}t \\ &= \int_0^T p(t) \sum_k (p_k + n_k)\phi_k(t)\,\mathrm{d}t \\ &= \int_0^T p(t)[p(t) + N_r(t)]\,\mathrm{d}t \end{aligned} \tag{7.28}$$

where $N_r(t)$ is the relevant noise part of $N(t)$ (see Appendix A or reference [14]). Since the irrelevant part of the noise $N_i(t)$ is orthogonal to the signal space (see Appendix A), adding this part of the noise to the relevant noise in the latter expression does not influence the result of the integration:

$$\begin{aligned}\sum_k p_k r_k &= \int_0^T p(t)[p(t) + N_r(t) + N_i(t)]\,\mathrm{d}t \\ &= \int_0^T p(t)[p(t) + N(t)]\,\mathrm{d}t = \int_0^T p(t)R(t)\,\mathrm{d}t \end{aligned} \tag{7.29}$$

The implementation of this operation is as follows. The received signal $R(t)$ is applied to a linear, time-invariant filter with the impulse response $p(T - t)$. The output of this filter is sampled at the end of the bit interval (at $t_0 = T$), and this sample value yields the statistic of Equation (7.29). This is a simple consequence of the convolution integral. The output signal of the filter is denoted by $Y(t)$, so that

$$Y(t) = R(t) * h(t) = R(t) * p(T - t) = \int_0^T R(\tau)h(t - \tau)\,\mathrm{d}\tau = \int_0^T R(\tau)p(\tau - t + T)\,\mathrm{d}\tau \tag{7.30}$$

At the sampling instant $t_0 = T$ the value of the signal at the output is

$$Y(T) = \int_0^T R(\tau)p(\tau)\,\mathrm{d}\tau \tag{7.31}$$

The detection process proceeds as indicated in Section 7.1.1; i.e. the sample value $Y(T)$ is compared to the threshold value. This threshold value D is found from Equations (7.27) and

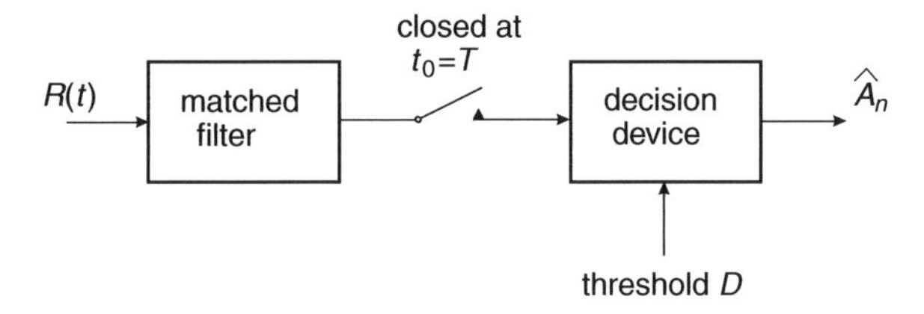

Figure 7.3 Optimal detector for binary signals

(7.7), and is implicitly determined by

$$\Lambda_0 = \exp\left(\frac{2D - E_p}{N_0}\right) \tag{7.32}$$

Hypothesis H_0 is chosen whenever $Y(T) < D$, whereas H_1 is chosen whenever $Y(T) \geq D$. In the special binary case where $P_0 = P_1 = \frac{1}{2}$, it follows that $D = E_p/2$. Note that in the case at hand the signal space will be one-dimensional.

The filter with the impulse response $h(t) = p(T - t)$ is called a matched filter, since the shape of its impulse response is matched to the pulse $p(t)$. The scheme of the detector is very simple; namely the signal is filtered by the matched filter and the output of this filter is sampled at the instant $t_0 = T$. If the sampled value is smaller than D then the detected bit is $\hat{A}_n = 0\,(H_0)$ and if the sampled value is larger than D then the receiver decides $\hat{A}_n = 1\,(H_1)$. This is represented schematically in Figure 7.3.

7.1.3 Detection of *M*-ary Signals in White Gaussian Noise

The situation of M-ary transmission is a generalization of the binary case. Instead of two different hypotheses and corresponding signals there are M different hypotheses, defined as

$$\begin{array}{ll} H_0: & R(t) = p_0(t) + N(t) \\ H_1: & R(t) = p_1(t) + N(t) \\ \cdot & \cdots \\ H_i: & R(t) = p_i(t) + N(t) \\ \cdot & \cdots \\ H_M: & R(t) = p_M(t) + N(t) \end{array} \tag{7.33}$$

As an example of this situation we mention FSK; in binary FSK we have $M = 2$.

To deal with the M-ary detection problem we do not use the likelihood ratio directly; in order to choose the maximum likely hypothesis we take a different approach. We turn to our fundamental criterion; namely the detector chooses the hypothesis that is most probable, given the received signal. The probabilities of the different hypotheses, given the received signal, are given by Equation (7.6). When selecting the hypothesis with the highest probability, the denominator $f_{\mathbf{r}}(\mathbf{r})$ may be ignored since it is common for all hypotheses. We are therefore looking for the hypothesis H_i, for which $P_i\, f_{\mathbf{r}}(\mathbf{r}\,|\,H_i)$ attains a maximum.

For a Gaussian noise probability density function this latter quantity is

$$\mathrm{P}_i f_{\mathbf{r}}(\mathbf{r} \mid \mathrm{H}_i) = \mathrm{P}_i \frac{1}{\sqrt{(\pi N_0)^K}} \exp\left[-\frac{1}{N_0}\sum_{k=1}^{K}(r_k - p_{k,i})^2\right], \quad i = 1, \ldots, M \tag{7.34}$$

with $p_{k,i}$ the kth element of $p_i(t)$ in the signal space; the summation over k actually represents the distance in signal space between the received signal and $p_i(t)$ and is called the distance metric. Since Equation (7.34) is a monotone-increasing function of $\mathbf{r}$, the decision may also be based on the selection of the largest value of the logarithm of expression (7.34). This means that we compare the different values of

$$\begin{aligned}
\ln[\mathrm{P}_i f_{\mathbf{r}}(\mathbf{r} \mid \mathrm{H}_i)] &= \ln \mathrm{P}_i - \frac{1}{N_0}\sum_{k=1}^{K}(r_k^2 + p_{k,i}^2 - 2 r_k p_{k,i}) - \ln\sqrt{(\pi N_0)^K} \\
&= \ln \mathrm{P}_i - \frac{1}{N_0}\sum_{k=1}^{K} r_k^2 - \frac{1}{N_0}\sum_{k=1}^{K} p_{k,i}^2 + \frac{2}{N_0}\sum_{k=1}^{K} r_k p_{k,i} - \ln\sqrt{(\pi N_0)^K}, \quad i = 1, \ldots, M
\end{aligned} \tag{7.35}$$

The second and fifth term on the right-hand side are common for all hypotheses, i.e. they do not depend on i, and thus they may be ignored in the decision process. We finally end up with the decision statistics

$$d_i = \frac{N_0}{2}\ln \mathrm{P}_i - \frac{E_i}{2} + \sum_{k=1}^{K} r_k p_{k,i}, \quad i = 1, \ldots, M \tag{7.36}$$

where E_i is the energy in the signal $p_i(t)$ (see Appendix A) and Equation (7.35) has been multiplied by $N_0/2$, which is allowed since it is a constant and does not influence the decision. For ease of notation we define

$$b_i \stackrel{\triangle}{=} \frac{N_0}{2}\ln \mathrm{P}_i - \frac{E_i}{2} \tag{7.37}$$

so that

$$d_i = b_i + \sum_{k=1}^{K} r_k p_{k,i} \tag{7.38}$$

Based on Equations (7.37) and (7.38) we can construct the optimum detector. It is shown in Figure 7.4. The received signal is filtered by a bank of matched filters, the ith filter being matched to the signal $p_i(t)$. The outputs of the filters are sampled and the result represents the last term of Equation (7.38). Next, the bias terms b_i given by Equation (7.37) are added to these outputs, as indicated in the figure. The resulting values d_i are applied to a circuit that selects the largest, thereby producing the detected symbol $\hat{A}$. This symbol is taken from the alphabet $\{A_1, \ldots, A_M\}$, the same set of symbols from which the transmitter selected its symbols.

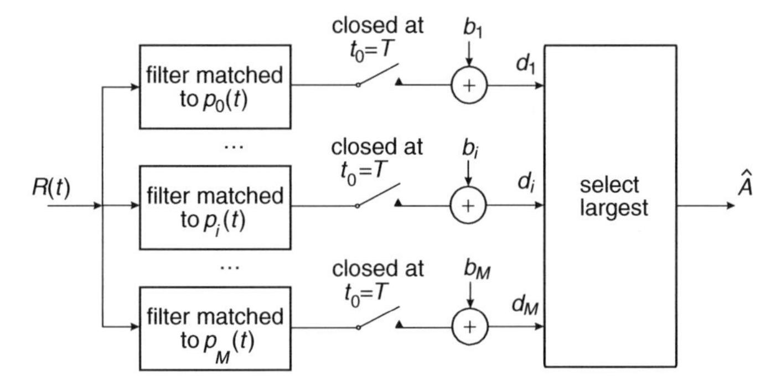

Figure 7.4 Optimal detector for M-ary signals where the symbols are mapped to different signals

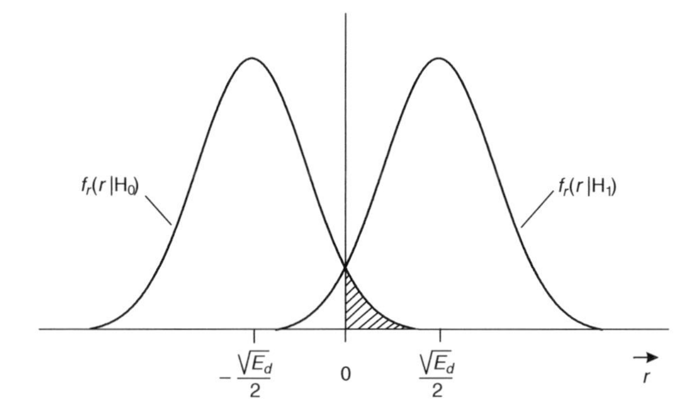

Figure 7.5 Conditional probability density functions in the binary case with the error probability indicated by the shaded area

In all these operations it is assumed that the shapes of the several signals $p_i(t)$ as well as their energy contents are known and fixed, and the prior probabilities P_i are known. It is evident that the bias terms may be omitted in case all prior probabilities are the same and all signals $p_i(t)$ carry the same energy.

Example 7.3:

As an example let us consider the detection of linearly independent binary signals $p_0(t)$ and $p_1(t)$ in white Gaussian noise. The signal space to describe this signal set is two-dimensional. However, it can be reduced to a one-dimensional signal space by converting to a simplex signal set (see Section A.5). The basis of this signal space is given by $\phi(t) = [p_0(t) - p_1(t)]/\sqrt{E_d}$, where E_d is the energy in the difference signal. The signal constellation is given by the coordinates $[-\sqrt{E_d}/2]$ and $[\sqrt{E_d}/2]$ and the distance between the signals amounts to $\sqrt{E_d}$. Superimposed on these signals is the noise with variance $N_0/2$ (see Appendix A, Equation (A.26)). We assume that the two hypotheses are equiprobable. The situation has been depicted in Figure 7.5. In this figure the two conditional probability density functions are presented. The error probability follows from Equation (7.17), and since the prior probabilities are equal we can conclude that

$$\mathrm{P_e} = \int_0^\infty f_r(r \mid \mathrm{H_0})\,\mathrm{d}r \tag{7.39}$$

In the figure the value of the error probability is indicated by the shaded area. Since the noise has been assumed to be Gaussian, this probability is written as

$$\mathrm{P_e} = \int_0^\infty \frac{1}{\sqrt{2\pi\frac{N_0}{2}}} \exp\left[-\frac{\left(x - \frac{\sqrt{E_\mathrm{d}}}{2}\right)^2}{N_0}\right] \mathrm{d}x \tag{7.40}$$

Introducing the change of integration variable

$$z \triangleq \frac{x - \frac{\sqrt{E_\mathrm{d}}}{2}}{\sqrt{\frac{N_0}{2}}} \Longrightarrow \mathrm{d}x = \sqrt{\frac{N_0}{2}}\,\mathrm{d}z \tag{7.41}$$

the error probability is written as

$$\mathrm{P_e} = \frac{1}{\sqrt{2\pi}} \int_{\sqrt{E_\mathrm{d}/(2N_0)}}^\infty \exp\left(-\frac{z^2}{2}\right) \mathrm{d}z \tag{7.42}$$

This expression is recognized as the well-known Q function (see Equation (7.19) and Appendix F). Finally, the error probability can be denoted as

$$\mathrm{P_e} = \mathrm{Q}\left(\sqrt{\frac{E_\mathrm{d}}{2N_0}}\right) \tag{7.43}$$

This is a rather general result that can be used for different binary transmission schemes, both for baseband and modulated signal formats. The conditions are that the noise is white, additive and Gaussian, and the prior probabilities are equal.

It is concluded that the error probability depends neither on the specific shapes of the received pulses nor on the signal set that has been chosen for the analysis, but only on the energy of the difference between the two pulses. Moreover, the error probability depends on the ratio E_d/N_0; this ratio can be interpreted as a signal-to-noise ratio, often expressed in dB (see Appendix B). In signal space the quantity E_d is interpreted as the squared distance of the signal points. The further the signals are apart in the signal space, the lower the error probability will be.

This specific example describes a situation that is often met in practice. Despite the fact that we have two linearly independent signals it suffices to provide the receiver with a single matched filter, namely a filter matched to the difference $p_0(t) - p_1(t)$, being the basis of the simplex signal set.

□

7.1.4 Decision Rules

1. *Maximum a posteriori probability (MAP) criterion.* Thus far the decision process was determined by the so-called posterior probabilities given by Equation (7.6). Therefore this rule is referred to as the maximum *a posteriori* probability (MAP) criterion.

2. *Maximum-likelihood (ML) criterion.* In order to apply the MAP criterion the prior probabilities should be known at the receiver. However, this is not always the case. In the absence of this knowledge it may be assumed that the prior probabilities for all the M signals are equal. A receiver based on this criterion is called a maximum-likelihood receiver.

3. *The Bayes criterion.* In our treatment we have considered the detection of binary data. In general, for signal detection a slightly different approach is used. The basics remain the same but the decision rules are different. This is due to the fact that in general the different detection probabilities are connected to certain costs. These costs are presented in a cost matrix

$$\mathbf{C} = \begin{pmatrix} C_{00} & C_{01} \\ C_{10} & C_{11} \end{pmatrix} \tag{7.44}$$

   where C_{ij} is the cost of H_i being detected when actually H_j is transmitted. In radar hypothesis H_1 corresponds to a target, whereas hypothesis H_0 corresponds to the absence of a target. Detecting a target when actually no target is present is called a false alarm, whereas detecting no target when actually one is there is called a miss. One can imagine that taking action on these mistakes can have severe consequences, which are differently weighed for the two different errors.

   The detection process can actually have four different outcomes, each of them associated with its own conditional probability. When applying the Bayes criterion the four different probabilities are multiplied by their corresponding cost factors, given by Equation (7.44). This results in the mean risk. The Bayes criterion minimizes this mean risk. For more details see reference [15].

4. *The minimax criterion.* The Bayes criterion uses the prior probabilities for minimizing the mean cost. When the detection process is based on wrong assumptions in this respect, the actual cost can be considerably higher than expected. When the probabilities are not known a good strategy is to minimize the maximum cost; i.e. whatever the prior probabilities in practice are, the mean cost can be guaranteed not to be larger than a certain value that can be calculated in advance. For further information on this subject see reference [15].

5. *The Neyman–Pearson criterion.* In radar detection the prior probabilities are often difficult to determine. In such situations it is meaningful to invoke the Neyman–Pearson criterion [15]. It maximizes the probability of detecting a target at a fixed false alarm probability. This criterion is widely used in radar detection.

7.2 FILTERS THAT MAXIMIZE THE SIGNAL-TO-NOISE RATIO

In this section we will derive a linear time-invariant filter that maximizes the signal-to-noise ratio when a known deterministic signal $x(t)$ is received and which is disturbed by additive noise. This maximum of the signal-to-noise ratio occurs at a specific, predetermined instant in time, the sampling instant. The noise need not be necessarily white or Gaussian. As we assumed in earlier sections, we will only assume it to be wide-sense stationary. The probability density function of the noise and its spectrum are allowed to have arbitrary shapes, provided they obey the conditions to be fulfilled for these specific functions.

Let us assume that the known deterministic signal may be Fourier transformed. The value of the output signal of the filter at the sampling instant t_0 is

$$y(t_0) = \frac{1}{2\pi}\int_{-\infty}^{\infty} X(\omega)H(\omega)\exp(\mathrm{j}\omega t_0)\,\mathrm{d}\omega \tag{7.45}$$

where $H(\omega)$ is the transfer function of the filter. Since the noise is supposed to be wide-sense stationary it follows from Equation (4.28) that the power of the noise output of the filter is

$$P_{N_0} = \mathrm{E}[N_0^2(t)] = \frac{1}{2\pi}\int_{-\infty}^{\infty} S_{NN}(\omega)|H(\omega)|^2\,\mathrm{d}\omega \tag{7.46}$$

with $S_{NN}(\omega)$ the spectrum of the input noise. The output signal power at the sampling instant is achieved by squaring Equation (7.45). Our goal is to find a value of $H(\omega)$ such that a maximum occurs for the signal-to-noise ratio defined as

$$\frac{S}{N} \triangleq \frac{|y(t_0)|^2}{P_{N_0}} = \frac{\left|\frac{1}{2\pi}\int_{-\infty}^{\infty} X(\omega)H(\omega)\exp(\mathrm{j}\omega t_0)\,\mathrm{d}\omega\right|^2}{\frac{1}{2\pi}\int_{-\infty}^{\infty} S_{NN}(\omega)|H(\omega)|^2\,\mathrm{d}\omega} \tag{7.47}$$

For this purpose we use the inequality of Schwarz. This inequality reads

$$\left|\int_{-\infty}^{\infty} A(\omega)B(\omega)\,\mathrm{d}\omega\right|^2 \leq \int_{-\infty}^{\infty} |A(\omega)|^2\,\mathrm{d}\omega \int_{-\infty}^{\infty} |B(\omega)|^2\,\mathrm{d}\omega \tag{7.48}$$

The equality holds if $B(\omega)$ is proportional to the complex conjugate of $A(\omega)$, i.e. if

$$A(\omega) = C\,B^*(\omega) \tag{7.49}$$

where C is an arbitrary real constant. With the substitutions

$$A(\omega) = H(\omega)\sqrt{S_{NN}(\omega)} \tag{7.50}$$

$$B(\omega) = \frac{X(\omega)\exp(\mathrm{j}\omega t_0)}{2\pi\sqrt{S_{NN}(\omega)}} \tag{7.51}$$

Equation (7.48) becomes

$$\left|\frac{1}{2\pi}\int_{-\infty}^{\infty} X(\omega)H(\omega)\exp(\mathrm{j}\omega t_0)\,\mathrm{d}\omega\right|^2 \leq \frac{1}{2\pi}\int_{-\infty}^{\infty} S_{NN}(\omega)|H(\omega)|^2\,\mathrm{d}\omega\,\frac{1}{2\pi}\int_{-\infty}^{\infty}\frac{|X(\omega)|^2}{S_{NN}(\omega)}\,\mathrm{d}\omega \tag{7.52}$$

From Equations (7.47) and (7.52) it follows that

$$\frac{S}{N} \leq \frac{1}{2\pi} \int_{-\infty}^{\infty} \frac{|X(\omega)|^2}{S_{NN}(\omega)} \, d\omega \tag{7.53}$$

It can be seen that in Equation (7.53) the equality holds if Equation (7.49) is satisfied. This means that the signal-to-noise ratio achieves its maximum value. From the sequel it will become clear that for the special case of white Gaussian noise the filter that maximizes the signal-to-noise ratio is the same as the matched filter that was derived in Section 7.1.2. This name is also used in the generalized case we are dealing with here. From Equations (7.49), (7.50) and (7.51) the following theorem holds.

Theorem 12

The matched filter has the transfer function (frequency domain description)

$$H_{\text{opt}}(\omega) = \frac{X^*(\omega)}{S_{NN}(\omega)} \exp(-j\omega t_0) \tag{7.54}$$

with $X(\omega)$ the Fourier transform of the input signal $x(t)$, $S_{NN}(\omega)$ the power spectral density function of the additive noise and t_0 the sampling instant.

We choose the constant C equal to 2π. The transfer function of the optimal filter appears to be proportional to the complex conjugate of the amplitude spectrum of the received signal $x(t)$. Furthermore, $H_{\text{opt}}(\omega)$ appears to be inversely proportional to the noise spectral density function. It is easily verified that an arbitrary value for the constant C may be chosen. From Equation (7.47) it follows that a constant factor in $H(\omega)$ does not affect the signal-to-noise ratio. In other words, in $H_{\text{opt}}(\omega)$ an arbitrary constant attenuation or gain may be inserted.

The sampling instant t_0 does not affect the amplitude of $H_{\text{opt}}(\omega)$ but only the phase $\exp(-j\omega t_0)$. In the time domain this means a delay over t_0. The value of t_0 may, as a rule, be chosen arbitrarily by the system designer and in this way may be used to guarantee a condition for realizability, namely causality.

The result we derived has a general validity; this means that it is also valid for white noise. In that case we make the substitution $S_{NN}(\omega) = N_0/2$. Once more, choosing a proper value for the constant C, we arrive at the following transfer function of the optimal filter:

$$H_{\text{opt}}(\omega) = X^*(\omega) \exp(-j\omega t_0) \tag{7.55}$$

This expression is easily transformed to the time domain.

Theorem 13

The matched filter for the signal $x(t)$ in white additive noise has the impulse response (time domain description)

$$h_{\text{opt}}(t) = x(t_0 - t) \tag{7.56}$$

with t_0 the sampling instant.

From Theorem 13 it follows that the impulse response of the optimal filter is found by shifting the input signal by t_0 to the left over the time axis and mirroring it with respect to $t = 0$. This time domain description offers the opportunity to guarantee causality by setting $h(t) = 0$ for $t < 0$.

Comparing the result of Equation (7.56) with the optimum filter found in Section 7.1.2, it is concluded that in both situations the optimal filters show the same impulse response. This may not surprise us, since in the case of Gaussian noise the maximum signal-to-noise ratio implies a minimum probability of error. From this we can conclude that the matched filter concept has a broader application than the considerations given in Section 7.1.2.

Once the impulse response of the optimal filter is known, the output response of this filter to the input signal $x(t)$ can be calculated. This is obtained by applying the well-known convolution integral

$$y(t) = \int_{-\infty}^{\infty} h_{\mathrm{opt}}(\tau)x(t-\tau)\,\mathrm{d}\tau = \int_{-\infty}^{\infty} x(t_0-\tau)x(t-\tau)\,\mathrm{d}\tau \tag{7.57}$$

At the decision instant t_0 the value of the output signal $y(t_0)$ equals the energy of the incoming signal till the moment t_0, multiplied by an arbitrary constant that may be introduced in $h_{\mathrm{opt}}(t)$.

The noise power at the output of the matched filter is

$$P_{N_0} = \frac{1}{2\pi}\frac{N_0}{2}\int_{-\infty}^{\infty} |H_{\mathrm{opt}}(\omega)|^2\,\mathrm{d}\omega = \frac{N_0}{2}\int_{-\infty}^{\infty} h_{\mathrm{opt}}^2(t)\,\mathrm{d}t \tag{7.58}$$

The last equality in this equation follows from Parseval's formula (see Appendix G or references [7] and [10]). However, since we found that the impulse response of the optimal filter is simply a mirrored version in time of the received signal (see Equation (7.56)) it is concluded that

$$P_{N_0} = \frac{N_0}{2}E_x \tag{7.59}$$

with E_x the energy content of the signal $x(t)$. From Equations (7.57) and (7.59) the signal-to-noise ratio at the output of the filter can be deduced.

Theorem 14

The signal-to-noise ratio at the output of the matched filter at the sampling instant is

$$\left(\frac{S}{N}\right)_{\max} \triangleq \frac{|y(t_0)|^2}{P_{N_0}} = \frac{2E_x}{N_0} \tag{7.60}$$

with E_x the energy content of the received signal and $N_0/2$ the spectral density of the additive white noise.

Although a method exists to generalize the theory of Sections 7.1.2 and 7.1.3 to include coloured noise, we will present a simpler alternative here. This alternative reduces the problem of coloured noise to that of white noise, for which we now know the solution, as

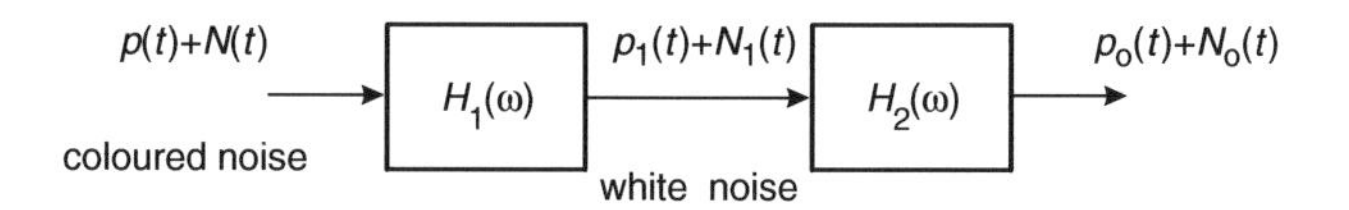

Figure 7.6 Matched filter for coloured noise

presented in the last paragraph. The basic idea is to insert a filter between the input and matched filter. The transfer function of this inserted filter is chosen such that the coloured input noise is transformed into white noise. The receiving filter scheme is as shown in Figure 7.6. It depicts the situation for hypothesis H_1, with the input $p(t) + N(t)$. The spectrum of $N(t)$ is assumed to be coloured. Based on what we want to achieve, the transfer function of the first filter should satisfy

$$|H_1(\omega)|^2 = \frac{1}{S_{NN}(\omega)} \tag{7.61}$$

By means of Equation (4.27) it is readily seen that the noise $N_1(t)$ at the output of this filter has a white spectral density. For this reason the filter is called a whitening filter. The spectrum of the signal $p_1(t)$ at the output of this filter can be written as

$$P_1(\omega) = P(\omega)H_1(\omega) \tag{7.62}$$

The problem therefore reduces to the white noise case in Theorem 13. The filter $H_2(\omega)$ has to be matched to the output of the filter $H_1(\omega)$ and thus reads

$$H_2(\omega) = \frac{P_1^*(\omega)}{S_{N_1N_1}(\omega)} \exp(-\mathrm{j}\omega t_0) = P^*(\omega)H_1^*(\omega)\exp(-\mathrm{j}\omega t_0) \tag{7.63}$$

In the second equation above we used the fact that $S_{N_1N_1}(\omega) = 1$, which follows from Equations (7.61) and (4.27). The matched filter for a known signal $p(t)$ disturbed by coloured noise is found when using Equations (7.61) and (7.63):

$$H(\omega) = H_1(\omega)H_2(\omega) = \frac{P^*(\omega)}{S_{NN}(\omega)} \exp(-\mathrm{j}\omega t_0) \tag{7.64}$$

It is concluded that the matched filter for a signal disturbed by coloured noise corresponds to the optimal filter from Equation (7.54).

Example 7.4:

Consider the signal

$$x(t) = \begin{cases} at, & 0 < t \leq T \\ 0, & \text{elsewhere} \end{cases} \tag{7.65}$$

This signal is shown in Figure 7.7(a). We want to characterize the matched filter for this signal when it is disturbed by white noise and to determine the maximum value of the

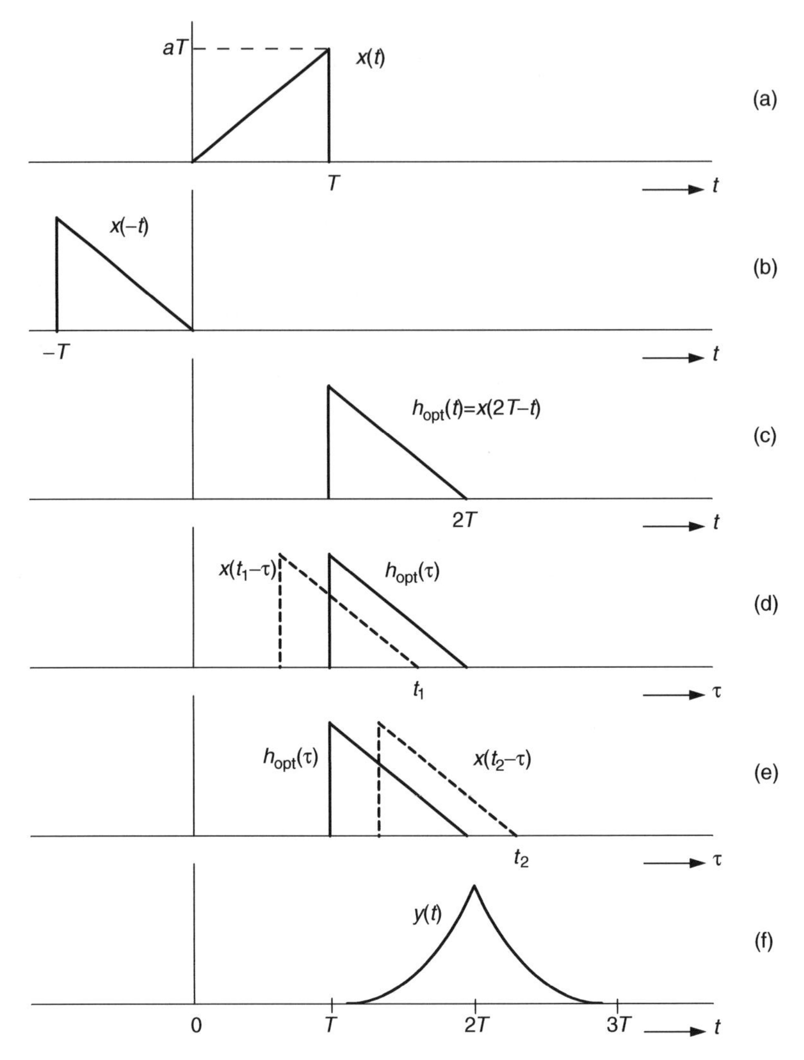

Figure 7.7 The different signals belonging to the example on a matched filter for the signal $x(t)$ disturbed by white noise

signal-to-noise ratio. The sampling instant is chosen as $t_0 = 2T$. In view of the simple description of the signal in the time domain it seems reasonable to do all the necessary calculations in the time domain. Illustrations of the different signals involved give a clear insight of the method. The signal $x(-t)$ is in Figure 7.7(b) and from this follows the optimal filter characterized by its impulse response $h_{\mathrm{opt}}(t)$, which is depicted in Figure 7.7(c). The maximum signal-to-noise ratio, occurring at the sampling instant t_0, is calculated as follows. The noise power follows from Equation (7.58) yielding

$$P_{N_0} = \frac{N_0}{2} \int_0^T a^2 t^2 \, \mathrm{d}t = \frac{N_0}{2} a^2 \frac{1}{3} t^3 \Big|_0^T = \frac{N_0 a^2 T^3}{6} \tag{7.66}$$

The signal value at $t = t_0$, using Equation (7.57), is

$$y(t_0) = \int_0^T a^2 t^2 \, dt = \frac{a^2 T^3}{3} \tag{7.67}$$

Using Equations (7.47), (7.66) and (7.67) the signal-to-noise ratio at the sampling instant t_0 is

$$\frac{S}{N} = \frac{y^2(t_0)}{P_{N_0}} = \frac{a^4 T^6/9}{N_0 a^2 T^3/6} = \frac{2a^2 T^3}{3N_0} \tag{7.68}$$

The output signal $y(t)$ follows from the convolution of $x(t)$ and $h_{\text{opt}}(t)$ as given by Equation (7.56). The convolution is

$$y(t) = \int_{-\infty}^{\infty} h_{\text{opt}}(\tau) x(t - \tau) \, d\tau \tag{7.69}$$

The various signals are shown in Figure 7.7. In Figure 7.7(d) the function $h_{\text{opt}}(\tau)$ has been drawn, together with $x(t - \tau)$ for $t = t_1$; the latter is shown as a dashed line. We distinguish two different situations, namely $t < t_0 = 2T$ and $t > t_0 = 2T$. In Figure 7.7(e) the latter case has been depicted for $t = t_2$. These pictures reveal that $y(t)$ has an even symmetry with respect to $t_0 = 2T$. That is why we confine ourselves to calculate $y(t)$ for $t \leq 2T$. Moreover, from the figures it is evident that $y(t)$ equals zero for $t < T$ and $t > 3T$. For $T \leq t \leq 2T$ we obtain (see Figure 7.7(d))

$$\begin{aligned} y(t) &= \int_T^t a(-\tau + 2T) a(-\tau + t) \, d\tau \\ &= a^2 \int_T^t (\tau^2 - t\tau - 2T\tau + 2Tt) \, d\tau \\ &= a^2 \left[\frac{1}{3}\tau^3 - \frac{t + 2T}{2}\tau^2 + 2Tt\tau \right]_T^t \\ &= a^2 \left[-\frac{1}{6}t^3 + Tt^2 - \frac{3}{2}T^2 t + \frac{2}{3}T^3 \right], \quad T \leq t \leq 2T \end{aligned} \tag{7.70}$$

The function $y(t)$ has been depicted in Figure 7.7(f). It is observed that the signal attains it maximum at $t = 2T$, the sampling instant.

□

The maximum of the output signal of a matched filter is always attained at t_0 and $y(t)$ always shows even symmetry with respect to $t = t_0$.

7.3 THE CORRELATION RECEIVER

In the former section we derived the linear time-invariant filter that maximizes the signal-to-noise ratio; it was called a matched filter. It can be used as a receiver filter prior to detection. It was shown in Section 7.1.2 that sampling and comparing the filtered signal with the proper threshold provides optimum detection of data signals in Gaussian noise. Besides matched filtering there is yet another method used to optimize the signal-to-noise ratio and which serves as an alternative for the matched filter. The method is called correlation reception.

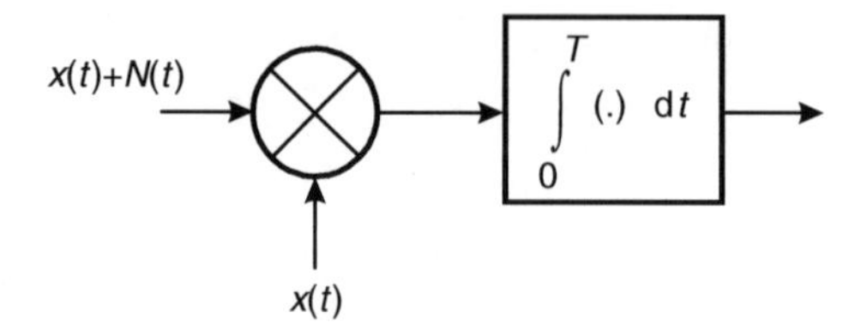

Figure 7.8 Scheme of the correlation receiver

The scheme of the correlation receiver is presented in Figure 7.8. In the receiver a synchronized replica of the information signal $x(t)$ has to be produced; this means that the signal must be known by the receiver. The incoming signal plus noise is multiplied by the locally generated $x(t)$ and the product is integrated. In the sequel we will show that the output of this system has the same signal-to-noise ratio as the matched filter.

For the derivation we assume that the pulse $x(t)$ extends from $t = 0$ to $t = T$. Moreover, the noise process $N(t)$ is supposed to be white with spectral density $N_0/2$. Since the integration is a linear operation, it is allowed to consider the two terms of the product separately. Applying only $x(t)$ to the input of the system of Figure 7.8 yields, at the output and at the sampling moment $t_0 = T$, the quantity

$$y(t_0) = \int_0^T x^2(t)\, \mathrm{d}t = E_x \tag{7.71}$$

where E_x is the energy in the pulse $x(t)$.

Next we calculate the power of the output noise as

$$\begin{aligned} P_{N_0} = \mathrm{E}[N_0^2(t)] &= \mathrm{E}\left[\int_0^T N(t)\, x(t)\, \mathrm{d}t \int_0^T N(\tau)\, x(\tau)\, \mathrm{d}\tau\right] \\ &= \mathrm{E}\left[\iint_0^T N(t)N(\tau)\, x(t)x(\tau)\, \mathrm{d}t\, \mathrm{d}\tau\right] \\ &= \iint_0^T \mathrm{E}[N(t)N(\tau)]\, x(t)x(\tau)\, \mathrm{d}t\, \mathrm{d}\tau \\ &= \iint_0^T \frac{N_0}{2}\delta(t-\tau)x(t)x(\tau)\, \mathrm{d}t\, \mathrm{d}\tau \\ &= \frac{N_0}{2}\int_0^T x^2(t)\, \mathrm{d}t = \frac{N_0}{2}E_x \end{aligned} \tag{7.72}$$

Then the signal-to-noise ratio is found from Equations (7.71) and (7.72) as

$$\frac{S}{N} = \frac{|y(t_0)|^2}{P_{N_0}} = \frac{E_x^2}{\frac{N_0}{2}E_x} = \frac{2E_x}{N_0} \tag{7.73}$$

This is exactly the same as Equation (7.60).

From the point of view of S/N, the matched filter receiver and the correlation receiver behave identically. However, for practical application it is of importance to keep in mind that there are crucial differences. The correlation receiver needs a synchronized replica of the known signal. If such a replica cannot be produced or if it is not exactly synchronized, the calculated signal-to-noise ratio will not be achieved, yielding a lower value. Synchronization is the main problem in using the correlation receiver. In many carrier-modulated systems it is nevertheless employed, since in such situations the phased–locked loop provides an excellent expedient for synchronization. The big advantage of the correlation receiver is the fact that all the time it produces, apart from the noise, the squared value of the signal. Together with the integrator this gives a continuously increasing value of the output signal, which makes the receiver quite invulnerable to deviations from the optimum sampling instant. This is in contrast to the matched filter receiver. If, for instance, the information signal changes its sign, as is the case in modulated signals, then the matched filter output changes as well. In this case a deviation from the optimum sampling instant can result in the wrong decision about the information bit. This is clearly demonstrated by the next example.

Example 7.5:

In the foregoing it was shown that the matched filter and the correlation receiver have equal performance as far as the signal-to-noise ratios at the sampling instant is concerned. In this example we compare the outputs of the two receivers when an ASK modulated data signal has to be received. It suffices to consider a single data pulse isolated in time. Such a signal is written as

$$x(t) = \begin{cases} A\cos(\omega_0 t), & 0 \le t < T \\ 0, & \text{elsewhere} \end{cases} \tag{7.74}$$

and where the symbol time T is an integer multiple of the period of the carrier frequency, i.e. $T = n \times 2\pi/\omega_0$ and n is integer. As the sampling instant we take $t_0 = T$. Then the matched filter output, for our purpose ignoring the noise, is found as

$$\begin{aligned}
y(t) &= A^2 \int_0^t \cos[\omega_0(t-\tau)]\cos[\omega_0(-\tau+T)]\,\mathrm{d}\tau \\
&= \frac{1}{2}A^2 \int_0^t \cos[\omega_0(t-T)] + \cos[\omega_0(-2\tau+t+T)]\,\mathrm{d}\tau \\
&= \frac{1}{2}A^2 \left\{ t\cos(\omega_0 t) - \frac{1}{2\omega_0}\sin[\omega_0(-2\tau+t+T)]\Big|_0^t \right\} \\
&= \frac{1}{2}A^2 \left[t\cos(\omega_0 t) + \frac{1}{\omega_0}\sin(\omega_0 t)\right], \quad 0 < t \le T
\end{aligned} \tag{7.75}$$

and

$$y(t) = \frac{1}{2}A^2 \left[(2T-t)\cos(\omega_0 t) - \frac{1}{\omega_0}\sin(\omega_0 t)\right], \quad T < t < 2T \tag{7.76}$$

For other values of t the response is zero. The total response is given in Figure 7.9, with parameter values of $A = T = 1$, $n = 4$ and $\omega_0 = 8\pi$. Note the oscillating character of the response, which changes its sign frequently.

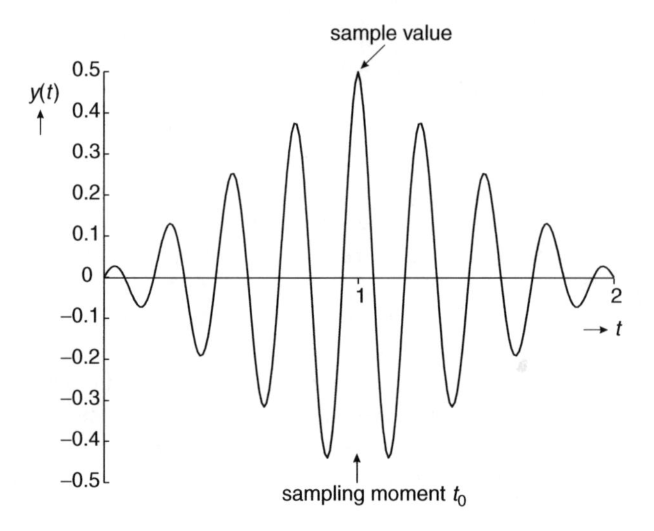

Figure 7.9 The response of the matched filter when driven by an ASK signal

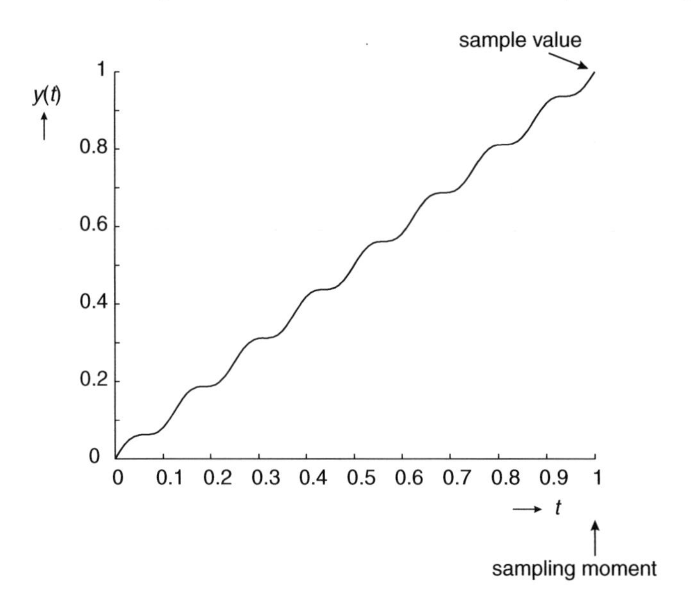

Figure 7.10 Response of the correlation receiver to an ASK signal

Next we consider the response of the correlation receiver to the same ASK signal as input. Now we find

$$\begin{aligned} y(t) &= A^2 \int_0^t \cos^2(\omega_0 \tau)\, \mathrm{d}\tau = \frac{1}{2} A^2 \int_0^t [1 + \cos(2\omega_0 \tau)]\, \mathrm{d}\tau \\ &= \frac{1}{2} A^2 t + \frac{A^2}{4\omega_0} \sin(2\omega_0 \tau) \bigg|_0^t = \frac{1}{2} A^2 t + \frac{A^2}{4\omega_0} \sin(2\omega_0 t), \quad 0 < t < T \end{aligned} \tag{7.77}$$

For negative values of t the response is zero and for $t > T$ the response depends on the specific design of the receiver. In an integrate-and-dump receiver the signal is sampled and subsequently the value of the integrator output is reset to zero. The response of Equation (7.77) is presented in Figure 7.10 for the same parameters as used to produce Figure 7.9. Note that in this case the response is continuously non-decreasing, so no change of sign occurs. This makes this type of receiver much less vulnerable to timing jitter of the sampler. However, a perfect synchronization is required instead.

□

7.4 FILTERS THAT MINIMIZE THE MEAN-SQUARED ERROR

Thus far it was assumed that the signal to be detected had a known shape. Now we proceed with signals that are not known in advance, but the shape of the signal itself has to be estimated. Moreover, we assume as in the former case that the signal is corrupted by additive noise. Although the signal is not known in the deterministic sense, some assumptions will be made about its stochastic properties; the same holds for the noise. In this section we make an estimate of the received signal in the mean-squared sense, i.e. we minimize the mean-squared error between an estimate of the signal based on available data consisting of signal plus noise and the actual signal itself. As far as the signal processing is concerned we confine the treatment to linear filtering.

Two different problems are considered.

1. In the first problem we assume that the data about the signal and noise are available for all times, so causality is ignored. We look for a linear time-invariant filtering that produces an optimum estimate for all times of the signal that is disturbed by the noise. This optimum linear filtering is called smoothing.
2. In the second approach causality is taken into account. We make an optimum estimate of future values of the signal based on observations in the past up until the present time. Once more the estimate uses linear time-invariant filtering and we call the filtering prediction.

7.4.1 The Wiener Filter Problem

Based on the description in the foregoing we consider a realization $S(t)$ of a wide-sense stationary process, called the signal. The signal is corrupted by the realization $N(t)$ of another wide-sense stationary process, called the noise. Furthermore, the signal and noise are supposed to be jointly wide-sense stationary. The noise is supposed to be added to the signal. To the input of the estimator the process

$$X(t) = S(t) + N(t) \tag{7.78}$$

is applied. When estimating the signal we base the estimate $\hat{S}(t+T)$ at some time $t+T$ on a linear filtering of the input data $X(t)$, i.e.

$$\hat{S}(t+T) = \int_a^b h(\tau)X(t-\tau)\,\mathrm{d}\tau \tag{7.79}$$

where $h(\tau)$ is the weighting function (equal to the impulse response of the linear time-invariant filter) and the integration limits a and b are to be determined later. Using Equation (7.79) the mean-squared error is defined as

$$e \triangleq \mathrm{E}\left[\{S(t+T)-\hat{S}(t+T)\}^2\right]=\mathrm{E}\left[\left\{S(t+T)-\int_a^b h(\tau)X(t-\tau)\,\mathrm{d}\tau\right\}^2\right] \tag{7.80}$$

Now the problem is to find the function $h(\tau)$ that minimizes the functional expression of Equation (7.80). In the minimization process the time shift T and integration interval are fixed; later on we will introduce certain restrictions to the shift and the integration interval, but for the time being they are arbitrary. The minimization problem can be solved by applying the calculus of variations [16]. According to this approach extreme values of the functional are achieved when the function $h(\tau)$ is replaced by $h(\tau)+\epsilon g(\tau)$, where $g(\tau)$ is an arbitrary function of the same class as $h(\tau)$. Next the functional is differentiated with respect to ϵ and the result equated to zero for $\epsilon=0$. Solving the resulting equation produces the function $h(\tau)$, which leads to the extreme value of the functional. In the next subsections we will apply this procedure to the problem at hand.

7.4.2 Smoothing

In the smoothing (or filtering) problem it is assumed that the data (or observation) $X(t)$ are known for the entire time axis $-\infty<t<\infty$. This means that there are no restrictions on the integration interval and we take $a\to-\infty$ and $b\to\infty$. Expanding Equation (7.80) yields

$$\begin{aligned} e=&\ \mathrm{E}[S^2(t+T)]+\mathrm{E}\left[\left(\int_{-\infty}^{\infty} h(\tau)X(t-\tau)\,\mathrm{d}\tau\right)^2\right]\\ &-\mathrm{E}\left[2S(t+T)\int_{-\infty}^{\infty} h(\tau)X(t-\tau)\,\mathrm{d}\tau)\right] \end{aligned} \tag{7.81}$$

Evaluating the expectations we obtain

$$\begin{aligned} e=&\ R_{SS}(0)+\iint_{-\infty}^{\infty} h(\tau)h(\rho)\mathrm{E}[X(t-\tau)X(t-\rho)]\,\mathrm{d}\tau\,\mathrm{d}\rho\\ &-2\int_{-\infty}^{\infty} h(\tau)\mathrm{E}[S(t+T)X(t-\tau)]\,\mathrm{d}\tau \end{aligned} \tag{7.82}$$

and further

$$e=R_{SS}(0)+\iint_{-\infty}^{\infty} h(\tau)h(\rho)R_{XX}(\tau-\rho)\,\mathrm{d}\tau\,\mathrm{d}\rho-2\int_{-\infty}^{\infty} h(\tau)R_{SX}(-\tau-T)\,\mathrm{d}\tau \tag{7.83}$$

According to the calculus of variations we replace $h(\tau)$ by $h(\tau) + \epsilon g(\tau)$ and obtain

$$e_\epsilon = R_{SS}(0) + \iint_{-\infty}^{\infty} [h(\tau) + \epsilon g(\tau)][h(\rho) + \epsilon g(\rho)] R_{XX}(\tau - \rho)\, \mathrm{d}\tau\, \mathrm{d}\rho$$
$$- 2\int_{-\infty}^{\infty} [h(\tau) + \epsilon g(\tau)] R_{SX}(-\tau - T)\, \mathrm{d}\tau \tag{7.84}$$

The procedure proceeds by setting

$$\left.\frac{\mathrm{d}e_\epsilon}{\mathrm{d}\epsilon}\right|_{\epsilon=0} = 0 \tag{7.85}$$

After some straightforward calculations this leads to the solution

$$R_{SX}(-\tau - T) = R_{XS}(\tau + T) = \int_{-\infty}^{\infty} h(\rho) R_{XX}(\tau - \rho)\, \mathrm{d}\rho, \quad -\infty < \tau < \infty \tag{7.86}$$

Since we assumed that the data are available over the entire time axis we can imagine that we apply this procedure on stored data. Moreover, in this case the integral in Equation (7.86) can be Fourier transformed as

$$S_{XS}(\omega) \exp(\mathrm{j}\omega T) = H(\omega)\, S_{XX}(\omega) \tag{7.87}$$

Hence we do not need to deal with the integral equation, which is now transformed into an algebraic equation. For the filtering problem we can set $T = 0$ and the optimum filter follows immediately:

$$H_{\mathrm{opt}}(\omega) = \frac{S_{XS}(\omega)}{S_{XX}(\omega)} \tag{7.88}$$

In the special case that the processes $S(t)$ and $N(t)$ are independent and at least one of these processes has zero mean, then the spectra can be written as

$$S_{XX}(\omega) = S_{SS}(\omega) + S_{NN}(\omega) \tag{7.89}$$
$$S_{XS}(\omega) = S_{SS}(\omega) \tag{7.90}$$

and as a consequence the optimum filter characteristic becomes

$$H_{\mathrm{opt}}(\omega) = \frac{S_{SS}(\omega)}{S_{SS}(\omega) + S_{NN}(\omega)} \tag{7.91}$$

Once we have the expression for the optimum filter the mean-squared error of the estimate can be calculated. For this purpose multiply both sides of Equation (7.86) by $h_{\mathrm{opt}}(\tau)$ and integrate over τ. This reveals that, apart from the minus sign, the second term of Equation (7.83) is half of the value of the third term, so that

$$e_{\min} = R_{SS}(0) - \int_{-\infty}^{\infty} h_{\mathrm{opt}}(\tau) R_{SX}(-\tau)\, \mathrm{d}\tau \tag{7.92}$$

If we define

$$\xi(t) \triangleq R_{SS}(t) - \int_{-\infty}^{\infty} h_{\text{opt}}(\tau) R_{SX}(t-\tau)\, d\tau \tag{7.93}$$

it is easy to see that

$$e_{\min} = \xi(0) \tag{7.94}$$

The Fourier transform of $\xi(t)$ is

$$S_{SS}(\omega) - H_{\text{opt}}(\omega)\, S_{SX}(\omega) = S_{SS}(\omega) - \frac{S_{SS}(\omega)\, S_{SX}(\omega)}{S_{XX}(\omega)} \tag{7.95}$$

Hence the minimum mean-squared error is

$$e_{\min} = \frac{1}{2\pi} \int_{-\infty}^{\infty} \left[S_{SS}(\omega) - \frac{S_{SS}(\omega)\, S_{SX}(\omega)}{S_{XX}(\omega)} \right] d\omega \tag{7.96}$$

When the special case of independence of the processes $S(t)$ and $N(t)$ is once more invoked, i.e. Equations (7.89) and (7.90) are inserted, then

$$e_{\min} = \frac{1}{2\pi} \int_{-\infty}^{\infty} \frac{S_{SS}(\omega)\, S_{NN}(\omega)}{S_{SS}(\omega) + S_{NN}(\omega)}\, d\omega \tag{7.97}$$

Example 7.6:

A wide-sense stationary process has a flat spectrum within a limited frequency band, i.e.

$$S_{SS}(\omega) = \begin{cases} S/2, & |\omega| \leq W \\ 0, & |\omega| > W \end{cases} \tag{7.98}$$

The noise is independent of $S(t)$ and has a white spectrum with a spectral density of $N_0/2$. In this case the optimum smoothing filter has the transfer function

$$H_{\text{opt}}(\omega) = \begin{cases} \dfrac{S}{S + N_0}, & |\omega| \leq W \\ 0, & |\omega| > W \end{cases} \tag{7.99}$$

This result can intuitively be understood; namely the signal spectrum is completely passed undistorted by the ideal lowpass filter of bandwidth W and the noise is removed outside the signal bandwidth. The estimation error is

$$e_{\min} = \frac{1}{2\pi} \int_{0}^{W} \frac{S N_0}{S + N_0}\, d\omega = \frac{W}{2\pi} \frac{S N_0}{S + N_0} = \frac{W}{2\pi} \frac{S}{S/N_0 + 1} \tag{7.100}$$

Interpreting S/N_0 as the signal-to-noise ratio it is observed that the error decreases with increasing signal-to-noise ratios. For a large signal-to-noise ratio the error equals the noise power that is passed by the filter.

□

The filtering of an observed signal as described by Equation (7.91) is also called signal restoration. It is the most obvious method when the observation is available as stored data. When this is not the case, but an incoming signal has to be processed in real time, then the filtering given in Equation (7.88) can be applied, provided that the delay is so large that virtually the whole filter response extends over the interval $-\infty < t < \infty$. Despite the real-time processing, a delay between the arrival of the signal $S(t)$ and the estimate of $S(t-T)$ should be allowed. In that situation the optimum filter has an extra factor of $\exp(-\mathrm{j}\omega T)$, which provides the delay, as follows from Equation (7.87). In general, a longer delay will reduce the estimation error, as long as the delay is shorter than the duration of the filter's impulse response $h(\tau)$.

7.4.3 Prediction

We now consider prediction based on the observation up to time t. Referring to Equation (7.79), we consider $\hat{S}(t+T)$ for positive values of T whereas $X(t)$ is only known up to t. Therefore the integral limits in Equation (7.79) are $a=-\infty$ and $b=t$. We introduce the causality of the filter's impulse response, given as

$$h(t) = 0, \quad \text{for } t < 0 \tag{7.101}$$

The general prediction problem is quite complicated [4]. Therefore we will confine the considerations here to the simplified case where the signal $S(t)$ is not disturbed by noise, i.e. now we take $N(t) \equiv 0$. This is called pure prediction. It is easy to verify that in this case Equation (7.86) is reduced to

$$R_{SS}(\tau + T) = \int_0^\infty h(\rho) R_{SS}(\tau - \rho)\, \mathrm{d}\rho, \quad \tau \geq 0 \tag{7.102}$$

This equation is known as the Wiener–Hopf integral equation. The solution is not as simple as in former cases. This is due to the fact that Equation (7.102) is only valid for $\tau \geq 0$; therefore we cannot use the Fourier transform to solve it. The restriction $\tau \geq 0$ follows from Equation (7.84). In the case at hand the impulse response $h(\tau)$ of the filter is supposed to be causal and the auxiliary function $g(\tau)$ should be of the same class. Consequently, the solution now is only valid for $\tau \geq 0$. For $\tau < 0$ the solution should be zero, and this should be guaranteed by the solution method.

Two approaches are possible for a solution. Firstly, a numerical solution can be invoked. For a fixed value of T the left-hand part of Equation (7.102) is sampled and the samples are collected in a vector. For the right-hand side $R_{SS}(\tau - \rho)$ is sampled for each value of τ. The different vectors, one for each τ, are collected in a matrix, which is multiplied by the unknown vector made up from the sampled values of $h(\rho)$. Finally, the solution is produced by matrix inversion.

Using an approximation we will also be able to solve it by means of the Laplace transform. Each function can arbitrarily be approximated by a rational function, the fraction of two polynomials. Let us suppose that the bilateral Laplace transform [7] of the autocorrelation function of $S(t)$ is written as a rational function, i.e.

$$S_{SS}(p) = \frac{A(p^2)}{B(p^2)} \tag{7.103}$$

Since the spectrum is an even function it can be written as a function of p^2. If we look at the positioning of zeros and poles in the complex p plane, it is revealed that this pattern is symmetrical with respect to the imaginary axis; i.e. if p_i is a root of $A(p^2)$ then $-p_i$ is a root as well. The same holds for $B(p^2)$. Therefore $S_{SS}(p)$ can be factored as

$$S_{SS}(p) = \frac{C(p)\,C(-p)}{D(p)\,D(-p)} = K(p)\,K(-p) \tag{7.104}$$

where $C(p)$ and $D(p)$ comprise all the roots in the left half-plane and $C(-p)$ and $D(-p)$ the roots in the right half-plane, respectively; $C(p)$ and $C(-p)$ contain the roots of $A^2(p)$ and $D(p)$ and $D(-p)$ those of $B^2(p)$. For the sake of convenient treatment we suppose that all roots are simple. Moreover, we define

$$K(p) \triangleq \frac{C(p)}{D(p)} \tag{7.105}$$

Both this function and its inverse are causal and realizable, since they are stable [7].

Let us now return to Equation (7.102), the integral equation to be solved. Rewrite it as

$$R_{SS}(\tau + T) = \int_0^{\infty} h(\rho) R_{SS}(\tau - \rho)\, \mathrm{d}\rho + f(\tau) \tag{7.106}$$

where $f(\tau)$ is a function that satisfies

$$f(\tau) = 0, \quad \text{for } \tau \geq 0 \tag{7.107}$$

i.e. $f(\tau)$ is anti-causal and analytic in the left-hand p plane ($\mathrm{Re}\{p\} < 0$). The Laplace transform of Equation (7.106) is

$$S_{SS}(p)\exp(pT) = S_{SS}(p)\,H(p) + F(p) \tag{7.108}$$

where $H(p)$ is the Laplace transform of $h(t)$ and $F(p)$ that of $f(t)$. Solving this equation for $H(p)$ yields

$$H(p) \triangleq \frac{N(p)}{M(p)} = \frac{\exp(pT)\,C(p)\,C(-p) - F(p)\,D(p)\,D(-p)}{C(p)\,C(-p)} \tag{7.109}$$

with the use of Equation (7.104). This function may only have roots in the left half-plane. If we select

$$F(p) = \frac{C(-p)}{D(-p)} \tag{7.110}$$

then Equation (7.109) becomes

$$H(p) = \frac{N(p)}{M(p)} = \frac{\exp(pT)\,C(p) - D(p)}{C(p)} \tag{7.111}$$

The choice given by Equation (7.110) guarantees that $f(t)$ is anti-causal, i.e. $f(t)=0$ for $t \geq 0$. Moreover, making

$$M(p) = C(p) \tag{7.112}$$

satisfies one condition on $H(p)$, namely that it is an analytic function in the right half-plane. Based on the numerator of Equation (7.111) we have to select $N(p)$; for that purpose the data of $D(p)$ can be used. We know that $D(p)$ has all its roots in the left half-plane, so if we select $N(p)$ such that its roots p_i coincide with those of $D(p)$ then the solution satisfies the condition that $N(p)$ is an analytic function in the right half-plane. This is achieved when the roots p_i are inserted in the numerator of Equation (7.111) to obtain

$$\exp(p_i T)\, C(p_i) = N(p_i) \tag{7.113}$$

for all the roots p_i of $D(p)$. Connecting the roots of the solution in this way to the polynomial $D(p)$ guarantees on the one hand that $N(p)$ is analytic in the right half-plane and on the other hand satisfies Equation (7.111). This completes the selection of the optimum $H(p)$.

Summarizing the method, we have to take the following steps:

1. Factor the spectral function

$$S_{SS}(p) = \frac{C(p)\, C(-p)}{D(p)\, D(-p)} \tag{7.114}$$

   where $C(p)$ and $D(p)$ comprise all the roots in the left half-plane and $C(-p)$ and $D(-p)$ the roots in the right half-plane, respectively.

2. The denominator of the optimum filter $H(p)$ has to be taken equal to $C(p)$.

3. Expand $K(p)$ into partial fractions:

$$K(p) = \frac{C(p)}{D(p)} = \frac{a_1}{p-p_1} + \cdots + \frac{a_n}{p-p_n} \tag{7.115}$$

   where p_i are the roots of $D(p)$.

4. Construct the modified polynomial

$$K_{\mathrm{m}}(p) = \exp(p_1 T)\frac{a_1}{p-p_1} + \cdots + \exp(p_n T)\frac{a_n}{p-p_n} \tag{7.116}$$

5. The optimum filter, described in the Laplace domain, then reads

$$H_{\mathrm{opt}}(p) = \frac{K_{\mathrm{m}}(p)\, D(p)}{C(p)} = \frac{N(p)}{C(p)} \tag{7.117}$$

Example 7.7:

Assume a process with the autocorrelation function

$$R_{SS}(\tau) = \exp(-\alpha|\tau|), \quad \alpha > 0 \tag{7.118}$$

Then from a Laplace transform table it follows that

$$S_{SS}(p) = \frac{2\alpha}{\alpha^2 - p^2} \tag{7.119}$$

which is factored into

$$S_{SS}(p) = \frac{\sqrt{2\alpha}}{\alpha + p} \frac{\sqrt{2\alpha}}{\alpha - p} \tag{7.120}$$

For the intermediate polynomial $K(p)$ it is found that

$$K(p) = \frac{\sqrt{2\alpha}}{\alpha + p} \tag{7.121}$$

Its constituting polynomials are

$$C(p) = \sqrt{2\alpha} \tag{7.122}$$

and

$$D(p) = \alpha + p \tag{7.123}$$

The polynomial $C(p)$ has no roots and the only root of $D(p)$ is $p_1 = -\alpha$. This produces

$$K_{\mathrm{m}}(p) = \frac{\sqrt{2\alpha}}{\alpha + p} \exp(-\alpha T) \tag{7.124}$$

so that finally for the optimum filter we find

$$H_{\mathrm{opt}}(p) = \frac{N(p)}{M(p)} = \frac{K_{\mathrm{m}}(p)\, D(p)}{C(p)} = \exp(-\alpha T) \tag{7.125}$$

and the corresponding impulse response is

$$h_{\mathrm{opt}}(t) = \exp(-\alpha T)\, \delta(t) \tag{7.126}$$

The minimum mean-squared error is given by substituting zero for the lower limit in the integral of Equation (7.92). This yields the error

$$e_{\min} = R_{SS}(0) - \int_0^\infty h(\tau) R_{SS}(-\tau)\, \mathrm{d}\tau = 1 - \exp(-\alpha T) \tag{7.127}$$

□

This result reflects what may be expected, namely the facts that the error is zero when $T = 0$, which is actually no prediction, and that the error increases with increasing values of T.

7.4.4 Discrete-Time Wiener Filtering

Discrete-Time Smoothing:

Once the Wiener filter for continuous processes has been analysed, the time-discrete version follows straightforwardly. Equation (7.86) is the general solution for describing the different situations considered in this section. Its time-discrete version when setting the delay to zero is

$$R_{XS}[n] = \sum_{m=-\infty}^{\infty} h[m]\, R_{XX}[n-m], \quad \text{for all } n \tag{7.128}$$

Since this equation is valid for all n it is easily solved by taking the z-transform of both sides:

$$\tilde{S}_{XS}(z) = \tilde{H}(z)\, \tilde{S}_{XX}(z) \tag{7.129}$$

or

$$\tilde{H}_{\text{opt}}(z) = \frac{\tilde{S}_{XS}(z)}{\tilde{S}_{XX}(z)} \tag{7.130}$$

The error follows from the time-discrete counterparts of Equation (7.92) or Equation (7.96).

If both $R_{XX}[n]$ and $R_{XS}[n]$ have finite extent, let us say $R_{XX}[n] = R_{XS}[n] = 0$ for $|n| > N$, and if the extent of $h[m]$ is limited to the same range, then Equation (7.128) can directly be solved in the time domain using matrix notation. For this case we define the $(2N+1) \times (2N+1)$ matrix as

$$\mathbf{R}_{XX} \triangleq \begin{bmatrix} R_{XX}[0] & R_{XX}[1] & R_{XX}[2] & \cdots & R_{XX}[N] & 0 & \cdots & 0 \\ R_{XX}[1] & R_{XX}[0] & R_{XX}[1] & \cdots & R_{XX}[N-1] & R_{XX}[N] & \cdots & 0 \\ \vdots & \vdots & \vdots & \vdots & \vdots & \vdots & \vdots & \vdots \\ 0 & 0 & R_{XX}[N] & \cdots & R_{XX}[0] & R_{XX}[1] & \cdots & 0 \\ \vdots & \vdots & \vdots & \vdots & \vdots & \vdots & \vdots & \vdots \\ 0 & 0 & 0 & \cdots & R_{XX}[N] & R_{XX}[N-1] & \cdots & R_{XX}[0] \end{bmatrix} \tag{7.131}$$

Moreover, we define the $(2N+1)$ element vectors as

$$\mathbf{R}_{XS}^{\mathrm{T}} \triangleq [R_{XS}[-N]\; R_{XS}[-N+1] \cdots R_{XS}[0] \cdots R_{XS}[N-1]\; R_{XS}[N]] \tag{7.132}$$

and

$$\mathbf{h}^{\mathrm{T}} \triangleq [h[-N]\; h[-N+1] \cdots h[0] \cdots h[N-1]\; h[N]] \tag{7.133}$$

where $\mathbf{R}_{XS}^{\mathrm{T}}$ and $\mathbf{h}^{\mathrm{T}}$ are the transposed vectors of the column vectors $\mathbf{R}_{XS}$ and $\mathbf{h}$, respectively.

By means of these definitions Equation (7.128) is rewritten as

$$\mathbf{R}_{XS} = \mathbf{R}_{XX} \cdot \mathbf{h} \tag{7.134}$$

with the solution for the discrete-time Wiener smoothing filter

$$\mathbf{h}_{\text{opt}} = \mathbf{R}_{XX}^{-1} \cdot \mathbf{R}_{XS} \tag{7.135}$$

This matrix description fits well in a modern numerical mathematical software package such as Matlab, which provides compact and efficient programming of matrices. Programs developed in Matlab can also be downloaded into DSPs, which is even more convenient.

Discrete-Time Prediction:

For the prediction problem a discrete-time version of the method presented in Subsection 7.4.3 can be developed (see reference [12]). However, using the time domain approach presented in the former paragraph, it is very easy to include noise; i.e. there is no need to limit the treatment to pure prediction.

Once more we start from the discrete-time version of Equation (7.86), which is now written as

$$R_{XS}[n+K] = \sum_{m=0}^{\infty} h[m]\, R_{XX}[n-m], \quad \text{for all } n \tag{7.136}$$

since the filter should be causal, i.e. $h[m] = 0$ for $m < 0$. Comparing this equation with Equation (7.128) reveals that they are quite similar. There is a time shift in R_{XS} and a difference in the range of $h[m]$. For the rest the equations are the same. This means that the solution is also the same, provided that the matrix $\mathbf{R}_{XX}$ and the vectors $\mathbf{R}_{XS}^{\mathrm{T}}$ and $\mathbf{h}^{\mathrm{T}}$ are accordingly redefined. They become the $(2N+1) \times (N+1)$ matrix

$$\mathbf{R}_{XX} \triangleq \begin{bmatrix} R_{XX}[N] & 0 & \cdots & 0 \\ R_{XX}[N-1] & R_{XX}[N] & \cdots & 0 \\ \vdots & \vdots & \vdots & \vdots \\ R_{XX}[0] & R_{XX}[1] & \cdots & R_{XX}[N] \\ \vdots & \vdots & \vdots & \vdots \\ R_{XX}[N] & R_{XX}[N-1] & \cdots & R_{XX}[0] \end{bmatrix} \tag{7.137}$$

and the $(2N+1)$ element vector

$$\mathbf{R}_{XS}^{\mathrm{T}} \triangleq [R_{XS}[-N+K]\; R_{XS}[-N+1+K] \cdots R_{XS}[N-1+K]\; R_{XS}[N+K]] \tag{7.138}$$

respectively, and the $(N+1)$ element vector

$$\mathbf{h}^{\mathrm{T}} \triangleq [h[0]\; h[1] \cdots h[N-1]\; h[N]] \tag{7.139}$$

The estimation error follows from the discrete-time version of Equation (7.92), which is

$$e = R_{SS}[0] - \sum_{n=0}^{N} h[n]\, R_{SX}[-n] \tag{7.140}$$

When the noise $N[n]$ has zero mean and $S[n]$ and $N[n]$ are independent, they are orthogonal. This simplifies the cross-correlation of R_{SX} to R_{SS}.

7.5 SUMMARY

The optimal detection of binary signals disturbed by noise has been considered. The problem is reduced to hypothesis testing. When the noise has a Gaussian probability density function, we arrive at a special form of linear filtering, the so-called matched filtering. The optimum receiver for binary data signals disturbed by additive wide-sense stationary Gaussian noise consists of a matched filter followed by a sampler and a decision device.

Moreover, the matched filter can also be applied in situations where the noise (not necessarily Gaussian) has to be suppressed maximally compared to the signal value at a specific moment in time, called the sampling instant. Since the matched filter is in fact a linear time-invariant filter and the input noise is supposed to be wide-sense stationary, this means that the output noise variance is constant, i.e. independent of time, and that the signal attains its maximum value at the sampling instant. The name matched filter is connected to the fact that the filter characteristic (let it be described in the time or in the frequency domain) is determined by (matched to) both the shape of the received signal and the power spectral density of the disturbing noise.

Finally, filters that minimize the mean-squared estimation error (Wiener filters) have been derived. They can be used for smoothing of stored data or portions of a random signal that arrived in the past. In addition, filters that produce an optimal prediction of future signal values have been described. Such filters are derived both for continuous processes and discrete-time processes.

7.6 PROBLEMS

7.1 The input $R = P + N$ is applied to a detector. The random variable P represents the information and is selected from $P \in \{+1, -0.5\}$ and the selection occurs with the probabilities $\mathrm{P}(P = +1) = \frac{1}{4}$ and $\mathrm{P}(P = -0.5) = \frac{3}{4}$. The noise N has a triangular distribution $f_N(n) = \mathrm{tri}(n)$.

(a) Make a sketch of the weighted (by the prior probabilities) conditional distribution functions.

(b) Determine the optimal decision regions.

(c) Calculate the minimum error probability.

7.2 Consider a signal detector with input $R = P + N$. The random variable P is the information and is selected from $P \in \{+A, -A\}$, with A a constant, and this selection occurs with equal probability. The noise N is characterized by the Laplacian probability density function

$$f_N(n) = \frac{1}{\sigma\sqrt{2}} \exp\left(-\frac{\sqrt{2}\,|n|}{\sigma}\right)$$

(a) Sketch the conditional probability density functions and determine the decision regions, without making a calculation.

(b) Consider the minimum probability of error receiver. Derive the probability of error for this receiver as a function of the parameters A and σ.

(c) Determine the variance of the noise.

(d) Defining an appropriate signal-to-noise ratio S/N, determine the S/N to achieve an error probability of 10^{-5}.

7.3 The M-ary PSK (phase shift keying) signal is defined as

$$p(t) = A\cos\left\{\omega_0 t + (i-1)\frac{2\pi}{M}\right\}, \quad i = 1, 2, \ldots, M, \quad \text{for } 0 \leq t \leq T$$

where A and ω_0 are constants representing the carrier amplitude and frequency, respectively, and i is randomly selected depending on the codeword to be transmitted. In Appendix A this signal is called a multiphase signal. This signal is disturbed by wide-sense stationary white Gaussian noise with spectral density $N_0/2$.

(a) Make a picture of the signal constellation in the signal space for $M = 8$.

(b) Determine the decision regions and indicate them in the picture.

(c) Calculate the symbol error probability (i.e. the probability that a codeword is detected in error) for large values of the signal-to-noise ratio; assume, among others, that this error rate is dominated by transitions to nearest neighbours in the signal constellation. Express this error probability in terms of M, the mean energy in the codewords and the noise spectral density.

Hint: use Equation (5.65) for the probability density function of the phase.

7.4 A filter matched to the signal

$$x(t) = \begin{cases} A\left(1 - \frac{|t|}{T}\right), & 0 < |t| < T \\ 0, & \text{elsewhere} \end{cases}$$

has to be realized. The signal is disturbed by noise with the power spectral density

$$S_{NN}(\omega) = \frac{W}{W^2 + \omega^2}$$

with A, T and W positive, real constants.

(a) Determine the Fourier transform of $x(t)$.

(b) Determine the transfer function $H_{\text{opt}}(\omega)$ of the matched filter.

(c) Calculate the impulse response $h_{\text{opt}}(t)$. Sketch $h_{\text{opt}}(t)$.

(d) Is there any value of t_0 for which the filter becomes causal? If so, what is that value?

7.5 The signal $x(t) = \mathrm{u}(t)\exp(-Wt)$, with $W > 0$ and real, is applied to a filter together with white noise that has a spectral density of $N_0/2$.

(a) Calculate the transfer function of the filter that is matched to $x(t)$.

(b) Determine the impulse response of this filter. Make a sketch of it.

(c) Is there a value of t_0 for which the filter becomes causal?

(d) Calculate the maximum signal-to-noise ratio at the output.

7.6 In a frequency domain description as given in Equation (7.54) the causality of the matched filter cannot be guaranteed. Using Equation (7.54) show that for a matched filter for a signal disturbed by (coloured) noise the following integral equation gives a time domain description:

$$\int_{-\infty}^{\infty} h_{\mathrm{opt}}(\xi) R_{NN}(t-\xi)\,\mathrm{d}\xi = x(t_0 - t)$$

where the causality of the filter can now be guaranteed by setting the lower bound of the integral equal to zero.

7.7 A pulse

$$x(t) = \begin{cases} A\cos(\pi t/T), & |t| \le T/2 \\ 0, & |t| > T/2 \end{cases}$$

is added to white noise $N(t)$ with spectral density $N_0/2$.

Find $(S/N)_{\max}$ for a filter matched to $x(t)$ and $N(t)$.

7.8 The signal $x(t)$ is defined as

$$x(t) = \begin{cases} A, & 0 \le t < T \\ 0, & \text{elsewhere} \end{cases}$$

This signal is disturbed by wide-sense stationary white noise with spectral density $N_0/2$.

(a) Make a sketch of $x(t)$.

(b) Sketch the impulse response of the matched filter if the sampling moment is $t_0 = T$.

(c) Sketch the output signal $y(t)$ of the filter.

(d) Calculate the signal-to-noise ratio at the filter output at $t = t_0$.

(e) Show that the filter given in Problem 4.3 realizes the matched filter of this signal in white noise.

7.9 The signal $x(t)$ is defined as

$$x(t) = \begin{cases} \exp(-t/\alpha), & 0 \le t < \alpha \\ 0, & \text{elsewhere} \end{cases}$$

This signal is disturbed by wide-sense stationary white noise with spectral density $N_0/2$.

(a) Make a sketch of $x(t)$.

(b) Sketch the impulse response of the matched filter if the sampling moment is $t_0 = 3\alpha$.

(c) Calculate and sketch the output signal $y(t)$ of the filter.

(d) Calculate the signal-to-noise ratio at the filter output at $t = t_0$.

7.10 The signal $x(t)$ is defined as

$$x(t) = \begin{cases} A, & 0 \leq t < T/2 \\ -A, & T/2 \leq t < T \\ 0, & \text{elsewhere} \end{cases}$$

This signal is disturbed by wide-sense stationary white noise with spectral density $N_0/2$.

(a) Make a sketch of $x(t)$.

(b) Sketch the impulse response of the matched filter if the sampling moment is $t_0 = T$.

(c) Sketch the output signal $y(t)$ of the filter.

(d) Calculate the signal-to-noise ratio at the filter output at $t = t_0$ and compare this with the outcome of Problem 7.8.

7.11 The signal $p(t)$ is defined as

$$p(t) = \begin{cases} A, & 0 \leq t \leq T \\ 0, & \text{elsewhere} \end{cases}$$

This signal is received twice (this is called 'diversity') in the following ways:

$$R_1(t) = p(t) + N_1(t)$$

and

$$R_2(t) = \alpha\, p(t) + N_2(t)$$

where α is a constant. The noise processes $N_1(t)$ and $N_2(t)$ are wide-sense stationary, white, independent processes and they both have the spectral density of $N_0/2$. The signals $R_1(t)$ and $R_2(t)$ are received by means of matched filters. The outputs of the matched filters are sampled at $t_0 = T$.

(a) Sketch the impulse responses of the two matched filters.

(b) Calculate the signal-to-noise ratios at the sample moments of the two individual receivers.

(c) By means of a new receiver design we produce the signal $R_3(t) = R_1(t) + \beta R_2(t)$, where β is a constant. Sketch the impulse response of the matched filter for $R_3(t)$ and calculate the signal-to-noise ratio at the sampling moment for this receiver.

(d) For what value of β will the signal-to-noise ratio from (c) attain its maximum value? (This is called 'maximum ratio combining'.)

(e) Compare the signal-to-noise ratios from (b) with the maximum value found at (d) and explain eventual differences.

7.12 For the signal $x(t)$ which is defined as

$$x(t) = \begin{cases} A, & 0 \leq t \leq T \\ 0, & \text{elsewhere} \end{cases}$$

a matched filter has to be designed. As an approximation an RC filter is selected with the transfer function

$$H(\omega) = \frac{1}{1 + j\omega\tau_0}$$

with $\tau_0 = RC$. The disturbing noise is wide-sense stationary and white with spectral density $N_0/2$.

(a) Calculate and sketch the output signal $y(t)$ of the RC filter.

(b) Calculate the maximum value of the signal-to-noise ratio at the output of the filter. Find the value of τ_0 that maximizes this signal-to-noise ratio.

Hint: use the Matlab command `fsolve` to solve the non-linear equation.

(c) Determine and sketch the output of the filter that is matched to the signal $x(t)$.

(d) Calculate the signal-to-noise ratio of the matched filter output and determine the difference with that of the RC filter.

7.13 A binary transmission system, where '1's and '0's are transmitted, is disturbed by wide-sense stationary additive white Gaussian noise with spectral density $N_0/2$. The '1's are mapped on to the signal $3p(t)$ and '0's on to $p(t)$, where

$$p(t) = \begin{cases} A, & 0 \leq t \leq T \\ 0, & \text{elsewhere} \end{cases}$$

In the binary received sequence those pulses will not overlap in time.

(a) Sketch $p(t)$.

(b) Sketch the impulse response of the filter that is matched to $p(t)$ and the noise, for the sampling moment $t_0 = T$.

(c) Sketch $y(t)$, the output of the filter when the input is $p(t)$.

(d) Determine and sketch the conditional probability density functions when receiving a '0' and a '1', respectively.

(e) Calculate the bit error probability when the prior probabilities for sending a '1' or a '0' are equal and independent.

7.14 A binary transmission system, where '1's and '0's are transmitted, is disturbed by wide-sense stationary additive white Gaussian noise with spectral density $N_0/2$. The

'1's are mapped on to the signal $p(t)$ and the '0's on to $-p(t)$, where

$$p(t) = \begin{cases} \cos(\pi t/T), & 0 \le t < T \\ 0, & \text{elsewhere} \end{cases}$$

In the binary received sequence those pulses will not overlap in time.

(a) Sketch $p(t)$.

(b) Sketch the impulse response of the filter that is matched to $p(t)$ and the noise, for the sampling moment $t_0 = T$.

(c) Calculate and plot $y(t)$, the output of the filter when the input is $p(t)$.

(d) Determine and sketch the conditional probability density functions when receiving a '0' or a '1'.

(e) Calculate the bit error probability when the prior probabilities for sending a '1' or a '0' are equal and independent.

7.15 Well-known digital modulation formats are ASK (amplitude shift keying) (see also Example 7.5) and PSK (phase shift keying). If we take one data symbol out of a sequence these formats are described by

$$X(t) = A[n]\cos(\omega_0 t), \quad 0 \le t < T$$

where the bit levels $A[n]$ are selected from $A[n] \in \{0, 1\}$ for ASK and from $A[n] \in \{-1, 1\}$ for PSK. The angular frequency ω_0 is constant and T is the bit time. Moreover, it is assumed that $T = n \times 2\pi/\omega_0$ with n an integer; i.e. the bit time is an integer number times the period of the carrier frequency. These signals are disturbed by wide-sense stationary white Gaussian noise with spectral density $N_0/2$. Assume that the signals are detected by a correlation receiver, i.e. the received signal plus noise is multiplied by $\cos(\omega_0 t)$ prior to detection.

(a) Sketch the signal constellations of ASK and PSK in signal space.

(b) Calculate the bit error rates, assuming the bits are equal probable. Express the error rates in terms of the mean energy in a bit and the spectral density of the noise.

7.16 An FSK (frequency shift keying) signal is defined as

$$X(t) = \begin{cases} A[n]\cos(\omega_1 t), & 0 \le t < T \\ \breve{A}[n]\cos(\omega_2 t), & 0 \le t < T \end{cases}$$

where the bit levels $A[n]$ are selected from $A[n] \in \{0, 1\}$ and $\breve{A}[n]$ is the negated value of $A[n]$, i.e. $\breve{A}[n] = 0$ if $A[n] = 1$ and $\breve{A}[n] = 1$ if $A[n] = 0$. The quantities ω_1 and ω_2 are constants and T is the bit time; the signal is disturbed by wide-sense stationary white Gaussian noise with spectral density $N_0/2$. As in Problem 7.15, the signal is detected by a correlation receiver.

(a) Sketch the structure of the correlation receiver.

(b) What are the conditions for an orthogonal signals space?

(c) Sketch the signal constellation in the orthogonal signal space.

(d) Calculate the bit error rate expressed in the mean energy in a bit and the noise spectral density. Compare the outcome with that of ASK (Problem 7.15) and explain the similarity or difference.

7.17 The PSK system presented in Problem 7.15 is called BPSK (binary PSK) or phase reversal keying (PRK). Now consider an alternative scheme where the two possible phase realizations do not have opposite phases but are shifted $\pi/2$ in phase.

(a) Sketch the signal constellation in signal space and indicate the decision regions.

(b) Calculate the minimum bit error probability and compare it with that of BPSK. Explain the difference.

(c) Sketch the correlation receiver for this signal.

(d) Has this scheme advantages with respect to BPSK? What are the disadvantages?

7.18 A signal $S(t)$ is observed in the middle of noise, i.e. $X(t) = S(t) + N(t)$. The signal $S(t)$ and the noise $N(t)$ are jointly wide-sense stationary. Design an optimum smoothing filter to estimate the derivative $S'(t)$ of the signal.

7.19 Derive the optimum prediction filter for a signal with spectral density

$$S_{SS}(\omega) = \frac{1}{1 + \omega^4}$$

8

Poisson Processes and Shot Noise

8.1 INTRODUCTION

Random point processes described in this chapter deal with a sequences of events, where both the time and the amplitude of the random variable is of a discrete nature. However, in contrast to the discrete-time processes dealt with so far, the samples are not equidistant in time, but the randomness is in the arrival times. In addition, the sample values may also be random. In this chapter we will not deal with the general description of random point processes but will confine the discussion to a special case, namely those processes where the number of events k, in fixed time intervals of length T, is described by a Poisson probability distribution

$$\mathrm{P}(X = k,\ T) = \frac{\exp(-\lambda T)\,(\lambda T)^k}{k!}, \quad k \geq 0, \text{integer} \tag{8.1}$$

In this equation we do not use the notation for the probability density function, since it is a discrete function and thus indeed a probability (denoted by $\mathrm{P}(\cdot)$) rather than a density, which is denoted by $f_X(x)$.

An integer-valued stochastic process is called a Poisson process if the following properties hold [1,5]:

1. The probability that k events occur in any arbitrary interval of length T is given by Equation (8.1).
2. The number of events that occur in any arbitrary time interval is independent of the number of events that occur in any other arbitrary non-overlapping time interval.

Many physical phenomena are accurately modelled by a Poisson process, such as the emission of photons from a laser source, the generation of hole/electron pairs in photodiodes by means of photons, the arrival of phone calls in an exchange, the number of jobs offered to a processor and the emission of electrons from a cathode.

In this chapter we will deal with the most important parameters of both Poisson processes with a constant expectation, called homogeneous Poisson processes, and processes with an

---

Introduction to Random Signals and Noise W. van Etten

expectation that is a function of time, called inhomogeneous Poisson processes. In both models, the amplitude of the physical phenomenon that is related to the events is constant, for instance the arrival of phone calls one at a time. Moreover, we will develop some theory about stochastic processes, as far as it relates to Poisson processes. Finally, we will consider Poisson impulse processes; these are processes where the amplitude of the event is a random variable as well. An example of such a process is the number of electrons produced in a photomultiplier tube, where the primary electrons (electrons that are directly generated by photon absorption) are subject to a random gain. A similar effect occurs in an avalanche photodiode [8].

8.2 THE POISSON DISTRIBUTION

8.2.1 The Characteristic Function

When dealing with Poisson and related processes, it appears that the characteristic function is a convenient tool to use in the calculation of important properties of the process, such as mean value, variance, etc.; this will become clear later on. The characteristic function of a random variable X is defined as

$$\Phi(u) \triangleq \mathrm{E}[\exp(\mathrm{j}uX)] = \int_{-\infty}^{\infty} f_X(x) \exp(\mathrm{j}ux)\,\mathrm{d}x \tag{8.2}$$

where $f_X(x)$ is the probability density function of X. Note that according to this definition the characteristic function is closely related to the Fourier transform of the probability density function, namely $\Phi(-u)$ is the Fourier transform of $f_X(x)$. The variable u is just a dummy variable and has no physical meaning. When X is a discrete variable with the possible realizations $\{X_k\}$, then the definition becomes

$$\Phi(u) \triangleq \sum_k \mathrm{P}(X = X_k) \exp(\mathrm{j}uX_k) \tag{8.3}$$

Sometimes it is useful to consider the logarithm of the characteristic function

$$\Psi(u) \triangleq \ln \Phi(u) \tag{8.4}$$

This function is called the second characteristic function of the random variable X. From Equation (8.2) it follows that

$$\Phi(0) = \int_{-\infty}^{\infty} f_X(x)\,\mathrm{d}x = 1 \tag{8.5}$$

so that

$$\Psi(0) = 0 \tag{8.6}$$

For the random variable $Y = aX + b$, with a and b as constants, it follows that

$$\Phi_Y(u) = \mathrm{E}[\exp(\mathrm{j}uY)] = \exp(\mathrm{j}ub)\,\Phi_X(au) \tag{8.7}$$

and

$$\Psi_Y(u) = \mathrm{j}b\,u + \Psi_X(au) \tag{8.8}$$

Example 8.1:

In this chapter we are primarily interested in the Poisson distribution given by Equation (8.1). For convenience we take $T = 1$ and for this distribution it can be seen that

$$\Phi(u) = \exp(-\lambda)\sum_{k=0}^{\infty}\exp(\mathrm{j}uk)\frac{\lambda^k}{k!} = \exp\{\lambda[\exp(\mathrm{j}u) - 1]\} \tag{8.9}$$

and

$$\Psi(u) = \lambda[\exp(\mathrm{j}u) - 1] \tag{8.10}$$

□

When on the other hand the characteristic function is known, the probability density function can be restored from it using the inverse Fourier integral transform

$$f_X(x) = \frac{1}{2\pi}\int_{-\infty}^{\infty}\Phi(u)\exp(-\mathrm{j}ux)\,\mathrm{d}u \tag{8.11}$$

The next example shows that defining the characteristic function by means of the Fourier transform of the probability density function is a powerful tool and can greatly simplify certain calculations.

Example 8.2:

Suppose that we have two independent Poisson distributed random variables X and Y with parameters λ_X and λ_Y, respectively. Moreover, it is assumed that we are interested to know what the probability density function of the sum $S = X + Y$ of the two variables is. It is well known [1] that the probability density function of the sum of two independent random variables is found by convolving the two probability density functions. As a rule this is a cumbersome operation, which is greatly simplified using the concept of the characteristic function; namely from Fourier theory it is known that convolution in one domain is equivalent to multiplication in the other domain. Using this property we conclude that

$$\begin{aligned}\Phi_S(u) &= \Phi_X(u)\,\Phi_Y(u) = \exp\{\lambda_X[\exp(\mathrm{j}u) - 1]\}\,\exp\{\lambda_Y[\exp(\mathrm{j}u) - 1]\}\\ &= \exp\{(\lambda_X + \lambda_Y)[\exp(\mathrm{j}u) - 1]\}\end{aligned} \tag{8.12}$$

Since this expression corresponds to the characteristic function of the Poisson distribution with parameter $\lambda_X + \lambda_Y$, we may conclude that the sum of two independent Poisson distributed random variables produces another Poisson distributed random variable of which the parameter is given by the sum of the parameters of the constituting random variables.

□

The moments of the random variable X are defined by

$$\mathrm{E}[X^n] \triangleq m_n \tag{8.13}$$

It follows, using Fourier transform theory, that a relationship between the derivatives of the characteristic function and these moments can be established. This relationship is found by expanding the exponential of the integrand of Equation (8.2) as follows:

$$\Phi(u) = \int_{-\infty}^{\infty} f_X(x)\left(1 + \mathrm{j}ux + \cdots + \frac{(\mathrm{j}ux)^n}{n!} + \cdots\right)\mathrm{d}x \tag{8.14}$$

Assuming that integration term by term is allowed, it follows that

$$\Phi(u) = 1 + \mathrm{j}m_1 u + \cdots + m_n \frac{(\mathrm{j}u)^n}{n!} + \cdots \tag{8.15}$$

From this equation the moments can immediately be identified as

$$\frac{\mathrm{d}^n \Phi(0)}{\mathrm{d}u^n} = \mathrm{j}^n m_n \tag{8.16}$$

The operations leading to Equation (8.16) are allowed if all the moments m_n exist and the series expansion of Equation (8.15) converges absolutely at $u = 0$. In this case $f_X(x)$ is uniquely determined by its moments m_n.

Sometimes it is more interesting to consider the central moments, e.g. the second central moment or variance. Then the preceding operations are applied to the random variable $X - \mathrm{E}[X]$, but for such an important central moment as the variance there is an alternative, as will be shown in the sequel.

8.2.2 Cumulants

Consider a probability distribution of which all the moments of arbitrary order exist. In the characteristic function ju is replaced by p and the function that results is called the moment generating function. The logarithm of this function becomes

$$\begin{aligned}\Psi(p) &= \ln \Phi(p) \\ &= \ln \int f_X(x)\left(1 + xp + \frac{x^2}{2!}p^2 + \cdots + \frac{x^n}{n!}p^n + \cdots\right)\mathrm{d}x \\ &= \ln\left(1 + m_1 p + \frac{m_2}{2!}p^2 + \cdots + \frac{m_n}{n!}p^n + \cdots\right)\end{aligned} \tag{8.17}$$

Now, expanding $\Psi(p)$ into a Taylor series about $p = 0$ and remembering that

$$\ln(1 + x) = x - \frac{x^2}{2} + \frac{x^3}{3} - \cdots \tag{8.18}$$

it is found that

$$\begin{aligned}\Psi(p) &= m_1 p + \frac{m_2}{2!}p^2 + \cdots - \frac{m_1^2}{2}p^2 + \cdots \\ &= m_1 p + \frac{1}{2}(m_2 - m_1^2)p^2 + \cdots \\ &= m_1 p + \frac{\sigma^2}{2}p^2 + \sum_{k=3}^{\infty}\frac{\gamma_k}{k!}p^k = \sum_{k=1}^{\infty}\frac{\gamma_k}{k!}p^k\end{aligned} \tag{8.19}$$

where σ^2 is the variance of X and γ_k is the kth cumulant or semi-invariant [1,15].

Example 8.3:

Based on Equations (8.10) and (8.19) the mean and variance of a Poisson distribution are easily established. For this distribution we have (see Equation (8.10))

$$\Psi(p) = \lambda[\exp(p) - 1] = \lambda\left(1 + p + \frac{p^2}{2} + \cdots - 1\right) = \lambda p + \frac{\lambda}{2}p^2 + \cdots \tag{8.20}$$

Comparing this expression with Equation (8.19) and equating term by term reveals that for the Poisson distribution both the mean value and the variance equal the parameter λ.

□

8.2.3 Interarrival Time and Waiting Time

For such problems as queuing, it is of importance to know the probability density function of the time that elapses between two events; this time is called the interarrival time. Suppose that an event took place at time t; then the probability that the random waiting time W is greater than some fixed value w represents the probability that no event occurs in the time interval $\{t,\ t + w\}$, or

$$\mathrm{P}(W > w) = \mathrm{P}(0, w) = \exp(-\lambda w), \quad w \geq 0 \tag{8.21}$$

and

$$\mathrm{P}(W \leq w) = \begin{cases} 1 - \exp(-\lambda w), & w \geq 0 \\ 0, & w < 0 \end{cases} \tag{8.22}$$

Thus, the waiting time is an exponential random variable and its probability density function is written as

$$f_W(w) = \frac{\mathrm{d}\{\mathrm{P}(W \leq w)\}}{\mathrm{d}w} = \begin{cases} \lambda\exp(-\lambda w), & w \geq 0 \\ 0, & w < 0 \end{cases} \tag{8.23}$$

The mean waiting time has the value

$$\mathrm{E}[W] = \int_0^\infty \lambda w \exp(-\lambda w)\, \mathrm{d}w = \frac{1}{\lambda} \tag{8.24}$$

This result can also easily be understood when remembering that λ is the mean number of events per unit of time. Then the mean waiting time will be its inverse.

8.3 THE HOMOGENEOUS POISSON PROCESS

Let us consider an homogeneous Poisson process defined as the sum of impulses

$$X(t) = \sum_i \delta(t - t_i) \tag{8.25}$$

where the δ impulses $\delta(t - t_i)$ appear at random times $\{t_i\}$ governed by a Poisson distribution. This process can be used for modelling such physical processes as the detection of photons by a photodetector. Each time an arriving photon is detected, it causes a small impulse-shaped amount of current having an amplitude equal to the charge of an electron. This means that the photon arrival and thus the production of current impulses may be described by Equation (8.25), depicted as the input of Figure 8.1(b). Due to the travel time of the moving charges in the detector and the frequency-dependent load circuit (see Figure 8.1(a)), the response of the current through the load will have a different shape, for instance as given in Figure 8.1(b). The shape of this response is called $h(t)$, being

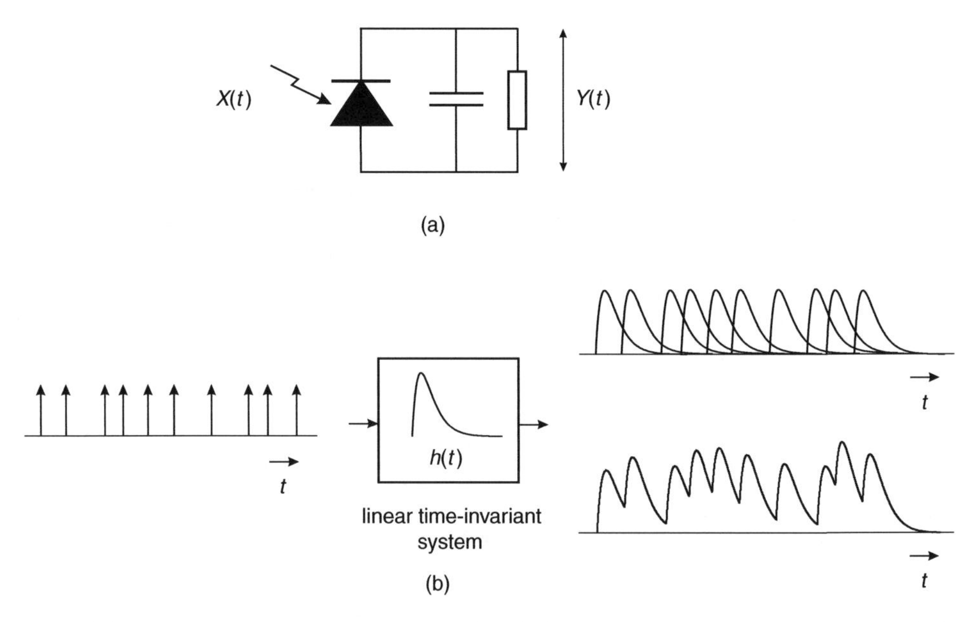

Figure 8.1 (a) Photodetector circuit with load and (b) corresponding Poisson process and shot noise process

the impulse response of the circuit. The total voltage across the load is described by the process

$$Y(t) = \sum_i h(t - t_i) \tag{8.26}$$

This process is called shot noise and even if the rate λ of the process is constant (in the example of the photodetector the constant amount of optical power arriving), the process $Y(t)$ will indeed show a noisy character, as has been depicted in the output of Figure 8.1(b). The upper curve at the output represents the individual responses, while the lower curve shows the sum of these pulses. Calculating the probability density function of this process is a difficult problem, but we will be able to calculate its mean value and autocorrelation function, and thus its spectrum as well.

8.3.1 Filtering of Homogeneous Poisson Processes and Shot Noise

In order to simplify the calculation of the mean value and the autocorrelation function of the shot noise, the time axis is subdivided into a sequence of consecutive small time intervals of equal width Δt. These time intervals are taken so short that $\lambda\,\Delta t \ll 1$. Next we define the random variable V_n such that for all integer values n

$$V_n = \begin{cases} 0, & \text{if no impulse occurs in the interval } n\,\Delta t < t < (n+1)\,\Delta t \\ 1, & \text{if one impulse occurs in the interval } n\,\Delta t < t < (n+1)\,\Delta t \end{cases} \tag{8.27}$$

In the sequel we shall neglect the probability that more than one impulse will occur in such a brief time interval as Δt. Following the definition of a Poisson process as given in the introduction to this chapter, it is concluded that the random variables V_n and V_m are independent if $n \neq m$. The probabilities of the two possible realizations of V_n read

$$P(V_n = 0) = \exp(-\lambda\,\Delta t) \approx 1 - \lambda\,\Delta t \tag{8.28}$$
$$P(V_n = 1) = \lambda\,\Delta t \exp(-\lambda\,\Delta t) \approx \lambda\,\Delta t \tag{8.29}$$

where the exact expressions are achieved by inserting $k = 0$ and $k = 1$, respectively, in Equation (8.1) and the approximations follow from the fact that $\lambda\,\Delta t \ll 1$. The expectation of the random variable V_n is

$$E[V_n] = 0 \times P(V_n = 0) + 1 \times P(V_n = 1) = P(V_n = 1) = \lambda\,\Delta t \tag{8.30}$$

Based on the foregoing it is easily revealed that the process $Y(t)$ is approximated by the process

$$\hat{Y}(t) = \sum_{n=-\infty}^{\infty} V_n\,h(t - n\,\Delta t) \tag{8.31}$$

The smaller the Δt the closer the approximation $\hat{Y}(t)$ will approach the shot noise process $Y(t)$; in the limit of Δt approaching zero, the processes will merge. The expectation of $\hat{Y}(t)$ is

$$\mathrm{E}[\hat{Y}(t)] = \sum_{n=-\infty}^{\infty} \mathrm{E}[V_n]\, h(t - n\,\Delta t) = \sum_{n=-\infty}^{\infty} \lambda\,\Delta t\, h(t - n\,\Delta t) \tag{8.32}$$

When Δt is made infinitesimally small then the summation is converted into an integral and the expected value of the shot noise process is obtained:

$$\mathrm{E}[Y(t)] = \lambda \int_{-\infty}^{\infty} h(t-\tau)\,\mathrm{d}\tau = \lambda \int_{-\infty}^{\infty} h(\tau)\,\mathrm{d}\tau \tag{8.33}$$

In order to gain more information about the shot noise process we will consider its characteristic function. This function is deduced in a straightforward way using the approximation $\hat{Y}(t)$:

$$\begin{aligned}
\Phi_{\hat{Y}}(u) &= \mathrm{E}[\exp\{ju\hat{Y}(t)\}] \\
&= \mathrm{E}\left[\exp\left\{ju \sum_{n=-\infty}^{\infty} V_n\, h(t - n\,\Delta t)\right\}\right] \\
&= \prod_{n=-\infty}^{\infty} \mathrm{E}[\exp\{juV_n\, h(t - n\,\Delta t)\}]
\end{aligned} \tag{8.34}$$

In Equation (8.34) the change from the summation in the exponential to the product of the expectation of the exponentials is allowed since the random variables V_n are independent. Invoking the law of total probability, the characteristic function of $\hat{Y}(t)$ is written as

$$\Phi_{\hat{Y}}(u) = \prod_{n=-\infty}^{\infty} \mathrm{P}(V_n = 0) \times 1 + \mathrm{P}(V_n = 1)\exp[juh(t - n\,\Delta t)] \tag{8.35}$$

and using Equations (8.28) and (8.29) gives

$$\begin{aligned}
\Phi_{\hat{Y}}(u) &= \prod_{n=-\infty}^{\infty} \{1 - \lambda\,\Delta t + \lambda\,\Delta t \exp[juh(t - n\,\Delta t)]\} \\
&= \prod_{n=-\infty}^{\infty} (1 + \lambda\,\Delta t\{\exp[juh(t - n\,\Delta t)] - 1\})
\end{aligned} \tag{8.36}$$

Now we use the approximation $1 + x \approx \exp(x)$ to proceed as

$$\begin{aligned}
\Phi_{\hat{Y}}(u) &\approx \prod_{n=-\infty}^{\infty} \exp(\lambda\,\Delta t\{\exp[juh(t - n\,\Delta t)] - 1\}) \\
&= \exp\left(\sum_{n=-\infty}^{\infty} \lambda\,\Delta t\{\exp[juh(t - n\,\Delta t)] - 1\}\right)
\end{aligned} \tag{8.37}$$

Once again Δt is made infinitesimally small so that the summation converts into an integral and we arrive at the characteristic function of the shot noise process $Y(t)$:

$$\begin{aligned}\Phi_Y(u) &= \exp\left(\lambda \int_{-\infty}^{\infty} \{\exp[juh(t-\tau)] - 1\}\, d\tau\right) \\ &= \exp\left(\lambda \int_{-\infty}^{\infty} \{\exp[juh(\tau)] - 1\}\, d\tau\right)\end{aligned} \tag{8.38}$$

From this result several features of the shot noise can be deduced. By inverse Fourier transforming it, we find the probability density function of $Y(t)$, but this is in general a difficult task. However, the mean and variance of the shot noise process follow immediately from the second characteristic function

$$\begin{aligned}\Psi_Y(u) = \ln \Phi_Y(u) &= \lambda \int_{-\infty}^{\infty} \{\exp[juh(\tau)] - 1\}\, d\tau \\ &= \lambda \int_{-\infty}^{\infty} \left[juh(\tau) - \tfrac{1}{2}u^2 h^2(\tau) + \cdots\right] d\tau\end{aligned} \tag{8.39}$$

and by invoking Equation (8.19).

□

Theorem 15

The homogeneous shot noise process has the mean value

$$\mathrm{E}[Y(t)] = \lambda \int_{-\infty}^{\infty} h(\tau)\, d\tau \tag{8.40}$$

and variance

$$\sigma_Y^2 = \lambda \int_{-\infty}^{\infty} h^2(\tau)\, d\tau \tag{8.41}$$

These two equations together are named Campbell's theorem.

Actually, the result of Equation (8.40) was earlier derived in Equation (8.33), but here we found this result in an alternative way. We emphasize that both the mean value and variance are proportional to the Poisson parameter λ. When the mean value of the shot noise process is interpreted as the signal and the variance as the noise, then it is concluded that this type of noise is not additive, as in the classical communication model, but multiplicative; i.e. the noise variance is proportional to the signal value and the signal-to-shot noise ratio is proportional to λ.

Example 8.4:

When we take a rectangular pulse for the impulse response of the linear time-invariant filter in Figure 8.1, according to

$$h(t) = \begin{cases} 1, & 0 \le t \le T \\ 0, & t < 0 \text{ and } t > T \end{cases} \tag{8.42}$$

then it is found that

$$\Phi_Y(u) = \exp\{\lambda T[\exp(\mathrm{j}u) - 1]\} \tag{8.43}$$

Comparing this result with Equation (8.9), it is concluded that in this case the output probability density function is a discrete one and gives the Poisson distribution of Equation (8.1). Actually, the filter gives as the output value at time t_s the number of Poisson impulses that arrived in the past T seconds, i.e. in the interval $t \in \{t_s - T,\ t_s\}$.

□

Next we want to calculate the autocorrelation function $R_Y(t_1, t_2)$ of the shot noise process and from that the power spectrum. For that purpose we define the joint characteristic function of two random variables X_1 and X_2 as

$$\Phi(u_1, u_2) \triangleq \mathrm{E}[\exp(\mathrm{j}u_1 X_1 + \mathrm{j}u_2 X_2)] = \iint_{-\infty}^{\infty} f_X(x_1, x_2) \exp(\mathrm{j}u_1 x_1 + \mathrm{j}u_2 x_2)\, \mathrm{d}x_1\, \mathrm{d}x_2 \tag{8.44}$$

Actually, this function is the two-dimensional Fourier transform of the joint probability density function of X_1 and X_2. In order to evaluate this function for the shot noise process $Y(t)$ we follow a similar procedure as before, i.e. we start with the approximating process $\hat{Y}(t)$:

$$\Phi_{\hat{Y}}(u_1, u_2) = \mathrm{E}[\exp\{\mathrm{j}u_1 \hat{Y}(t_1) + \mathrm{j}u_2 \hat{Y}(t_2)\}] \tag{8.45}$$

Elaborating this in a straightforward manner as before (see Equations (8.34) to (8.38)) we arrive at the joint characteristic function of $Y(t)$:

$$\Phi_Y(u_1, u_2) = \exp\left(\lambda \int_{-\infty}^{\infty} \{\exp[\mathrm{j}u_1 h(t_1 - \tau) + \mathrm{j}u_2 h(t_2 - \tau)] - 1\}\, \mathrm{d}\tau\right) \tag{8.46}$$

The second joint characteristic function reads

$$\Psi_Y(u_1, u_2) \triangleq \ln[\Phi_Y(u_1, u_2)] = \lambda \int_{-\infty}^{\infty} \{\exp[\mathrm{j}u_1 h(t_1 - \tau) + \mathrm{j}u_2 h(t_2 - \tau)] - 1\}\, \mathrm{d}\tau \tag{8.47}$$

In the sequel we shall show that by series expansion of this latter function, the autocorrelation function can be calculated. However, we will first show how moments of several orders are generated by the joint characteristic function. For that purpose we apply series expansion to both exponentials in Equation (8.44):

$$\begin{aligned}
\Phi(u_1, u_2) &= \iint_{-\infty}^{\infty} f_X(x_1, x_2) \sum_{n=0}^{\infty} \frac{(\mathrm{j}u_1 x_1)^n}{n!} \sum_{m=0}^{\infty} \frac{(\mathrm{j}u_2 x_2)^m}{m!}\, \mathrm{d}x_1\, \mathrm{d}x_2 \\
&= \iint_{-\infty}^{\infty} f_X(x_1, x_2) \left(1 + \mathrm{j}u_1 x_1 - \frac{u_1^2 x_1^2}{2} + \cdots\right)\left(1 + \mathrm{j}u_2 x_2 - \frac{u_2^2 x_2^2}{2} + \cdots\right) \mathrm{d}x_1\, \mathrm{d}x_2 \\
&= 1 + \mathrm{j}u_1 \mathrm{E}[X_1] + \mathrm{j}u_2 \mathrm{E}[X_2] - u_1 u_2 \mathrm{E}[X_1 X_2] + \cdots
\end{aligned} \tag{8.48}$$

From this equation it is observed that the term with $\mathrm{j}u_1$ comprises $\mathrm{E}[X_1]$, the term with $\mathrm{j}u_2$ comprises $\mathrm{E}[X_2]$, the term with $-u_1 u_2$ comprises $\mathrm{E}[X_1 X_2]$, etc. In fact, series expansion of the joint characteristic function generates all arbitrary moments as follows:

$$\Phi(u_1, u_2) = \sum_{n=0}^{\infty} \sum_{m=0}^{\infty} \mathrm{E}[X_1^n X_2^m] \frac{(\mathrm{j}u_1)^n (\mathrm{j}u_2)^m}{n!m!} \tag{8.49}$$

Since the characteristic function given by Equation (8.46) contains a double exponential, producing a series expansion is intractable. Therefore we make a series expansion of the second joint characteristic function of Equation (8.47) and identify from that expansion the second-order moment $\mathrm{E}[Y(t_1)Y(t_2)]$ we are looking for. This expansion is once again based on the series expansion of the logarithm $\ln(1 + x) = x - x^2/2 + \cdots$:

$$\begin{aligned}\Psi(u_1, u_2) &= \mathrm{j}u_1\mathrm{E}[X_1] + \mathrm{j}u_2\mathrm{E}[X_2] - u_1u_2\mathrm{E}[X_1X_2] - \tfrac{1}{2}u_1^2\mathrm{E}[X_1^2] - \tfrac{1}{2}u_2^2\mathrm{E}[X_2^2] + \cdots \\ &\quad - \tfrac{1}{2}(\mathrm{j}u_1\mathrm{E}[X_1] + \mathrm{j}u_2\mathrm{E}[X_2] - u_1u_2\mathrm{E}[X_1X_2] + \cdots)^2 + \cdots \\ &= \mathrm{j}u_1\mathrm{E}[X_1] + \mathrm{j}u_2\mathrm{E}[X_2] - u_1u_2\mathrm{E}[X_1X_2] - u_1^2\mathrm{E}[X_1^2] - u_2^2\mathrm{E}[X_2^2] + \cdots \\ &\quad - \tfrac{1}{2}(-u_1^2\mathrm{E}^2[X_1] - u_2^2\mathrm{E}^2[X_2] - 2u_1u_2\mathrm{E}[X_1]\mathrm{E}[X_2] + \cdots) + \cdots\end{aligned} \tag{8.50}$$

When looking at the term with u_1u_2, we discover that its coefficient reads $-(\mathrm{E}[X_1X_2] - \mathrm{E}[X_1]\mathrm{E}[X_2])$. Comparing this expression with Equation (2.65) and applying it to the process $Y(t)$ it is revealed that this coefficient equals the negative of the autocovariance function $C_{YY}(t_1, t_2)$. As we have already calculated the mean value of the process (see Equations (8.33) and (8.40)), we can easily obtain its autocorrelation function. To evaluate this function we expand Equation (8.47) in a similar way and look for the coefficient of the term with u_1u_2. This yields

$$C_{YY}(t_1, t_2) = \lambda \int_{-\infty}^{\infty} h(t_1 - \tau)\, h(t_2 - \tau)\, \mathrm{d}\tau = \lambda \int_{-\infty}^{\infty} h(\rho)\, h(t_2 - t_1 + \rho)\, \mathrm{d}\rho \tag{8.51}$$

We observe that this expression does not depend on the absolute time, but only on the difference $t_2 - t_1$. Since the mean was independent of time as well, it is concluded that the shot noise process is wide-sense stationary. Its autocorrelation function reads

$$R_{YY}(\tau) = C_{YY}(\tau) + \mathrm{E}^2[Y(t)] = \lambda \int_{-\infty}^{\infty} h(\rho)\, h(\tau + \rho)\, \mathrm{d}\rho + \lambda^2 \left[\int_{-\infty}^{\infty} h(\rho)\, \mathrm{d}\rho\right]^2 \tag{8.52}$$

From this we can immediately find the power spectrum by Fourier transforming the latter expression:

$$S_{YY}(\omega) = \lambda |H(\omega)|^2 + 2\pi\lambda^2 H^2(0)\delta(\omega) \tag{8.53}$$

It can be seen that the spectrum always comprises a d.c. component if this component is passed by the filter, i.e. if $H(0) \neq 0$. As a special case we consider the spectrum of the input process $X(t)$ that consists of a sequence of Poisson impulses as given by Equation (8.25). This spectrum is easily found by inserting $H(\omega) = 1$ in Equation (8.53). Then, apart from the δ function, the spectrum comprises a constant value, so this part of the spectrum behaves as white noise.

For large values of the Poisson parameter λ, the shot noise process approaches a Gaussian process [1]. This model is widely used for the shot noise generated by electronic components.

8.4 INHOMOGENEOUS POISSON PROCESSES

An inhomogeneous Poisson process is a Poisson process for which the parameter λ varies with time, i.e. $\lambda = \lambda(t)$. In order to derive the important properties of such a process we redo the calculations of the preceding section, where the parameter λ in Equations (8.28) to (8.37) is replaced by $\lambda(n\,\Delta t)$. The characteristic function then becomes

$$\Phi_Y(u) = \exp\left(\int_{-\infty}^{\infty} \lambda(\tau)\{\exp[\mathrm{j}uh(t-\tau)] - 1\}\,\mathrm{d}\tau\right) \tag{8.54}$$

Based on Equation (8.19), the mean and variance of this process follows immediately:

$$\mathrm{E}[Y(t)] = \int_{-\infty}^{\infty} \lambda(\tau)h(t-\tau)\,\mathrm{d}\tau \tag{8.55}$$

$$\sigma_Y^2 = \int_{-\infty}^{\infty} \lambda(\tau)h^2(t-\tau)\,\mathrm{d}\tau \tag{8.56}$$

Actually, these two equations are an extension of Campbell's theorem.

Without going into detail, the autocorrelation function of this process is easily found in a way similar to the procedure of calculating the second joint characteristic function in the preceding section. The second joint characteristic function of the inhomogeneous Poisson process is obtained as

$$\Psi_Y(u_1, u_2) = \int_{-\infty}^{\infty} \lambda(\rho)\{\exp[\mathrm{j}u_1 h(t_1-\rho) + \mathrm{j}u_2 h(t_2-\rho)] - 1\}\,\mathrm{d}\rho \tag{8.57}$$

and from this, once again, similarly to the preceding section, it follows that

$$\begin{aligned} R_{YY}(t_1, t_2) &= \int_{-\infty}^{\infty} \lambda(\rho)h(t_1-\rho)\,h(t_2-\rho)\,\mathrm{d}\rho \\ &\quad + \int_{-\infty}^{\infty} \lambda(\rho)h(t_1-\rho)\,\mathrm{d}\rho \int_{-\infty}^{\infty} \lambda(\rho)h(t_2-\rho)\,\mathrm{d}\rho \end{aligned} \tag{8.58}$$

Let us now suppose that the function $\lambda(t)$ is a stochastic process as well, which is independent of the Poisson process. Then the process $Y(t)$ is a doubly stochastic process [17]. Moreover, we assume that $\lambda(t)$ is a wide-sense stationary process. Then from Equation (8.55) it follows that the mean value of $Y(t)$ is independent of time. In the autocorrelation function we substitute $t_1 = t$ and $t_2 = t + \tau$; this yields

$$\begin{aligned} R_{YY}(t, t+\tau) &= \mathrm{E}[\lambda] \int_{-\infty}^{\infty} h(t-\rho)h(t+\tau-\rho)\,\mathrm{d}\rho \\ &\quad + \iint_{-\infty}^{\infty} R_{\lambda\lambda}(\rho_1-\rho_2)h(t-\rho_1)h(t+\tau-\rho_2)\,\mathrm{d}\rho_1\,\mathrm{d}\rho_2 \end{aligned} \tag{8.59}$$

In this equation $R_{\lambda\lambda}(\cdot)$ is the autocorrelation function of the process $\lambda(t)$. Further elaborating the Equation (8.59) gives

$$\begin{aligned} R_{YY}(t, t+\tau) &= \mathrm{E}[\lambda] \int_{-\infty}^{\infty} h(\xi)h(\xi+\tau)\,\mathrm{d}\xi \\ &+ \iint_{-\infty}^{\infty} R_{\lambda\lambda}(\tau+\xi_1-\xi_2)h(\xi_1)h(\xi_2)\,\mathrm{d}\xi_1\,\mathrm{d}\xi_2 \\ &= \mathrm{E}[\lambda]\; h(\tau) * h(-\tau) + R_{\lambda\lambda}(\tau) * h(\tau) * h(-\tau) \end{aligned} \tag{8.60}$$

Now it is concluded that the doubly stochastic process $Y(t)$ is wide-sense stationary as well. From this latter expression its power spectral density is easily revealed by Fourier transformation:

$$S_{YY}(\omega) = |H(\omega)|^2 \{\mathrm{E}[\lambda] + S_{\lambda\lambda}(\omega)\} \tag{8.61}$$

The interpretation of this expression is as follows. The first term reflects the filtering of the white shot noise spectrum and is proportional to the mean value of the information signal $\lambda(t)$. Note that $\mathrm{E}[\lambda(t)] = 0$ is meaningless from a physical point of view. The second term represents the filtering of the information signal.

An important physical situation where the theory in this section applies is a lightwave that is intensity modulated by an information signal. The detected current in the receiver is produced by photons arriving in the photodetector. This arrival of photons is then modelled as a doubly stochastic process with $\lambda(t)$ as the information signal [8].

8.5 THE RANDOM-PULSE PROCESS

Let us further extend the inhomogeneous process that was introduced in the preceding section. A Poisson impulse process consists of a sequence of δ functions

$$X(t) = \sum_i G_i \delta(t - t_i) \tag{8.62}$$

where the number of events per unit of time are governed by a Poisson distribution and $\{G_i\}$ are realizations of the random variable G; i.e. the amplitudes of the different impulses vary randomly and thus are subject to a random gain. In this section we again assume that the Poisson distribution may have a time-variant parameter $\lambda(t)$, which is supposed to be a wide-sense stationary stochastic process. Each of three random parameters involved is assumed to be independent of all the others. When filtering the process $X(t)$ we get the random-pulse process

$$Y(t) = \sum_i G_i h(t - t_i) \tag{8.63}$$

The properties of this process are again derived in a similar and straightforward way, as presented in Section 8.3.1. Throughout the entire derivation a third random variable is involved and the expectation over this variable has to be taken in addition. This leads to the following result for the characteristic function:

$$\Phi_Y(u) = \exp\left(\int_{-\infty}^{\infty} \lambda(\tau)\{\Phi_G[uh(t-\tau)] - 1\}\,\mathrm{d}\tau\right) \tag{8.64}$$

where $\Phi_G(u)$ is the characteristic function of G. From this another extension of Campbell's theorem follows:

$$\mathrm{E}[Y(t)] = \mathrm{E}[G] \int_{-\infty}^{\infty} \lambda(\tau)h(t-\tau)\,\mathrm{d}\tau \tag{8.65}$$

$$\sigma_Y^2 = \mathrm{E}[G^2] \int_{-\infty}^{\infty} \lambda(\tau)h^2(t-\tau)\,\mathrm{d}\tau \tag{8.66}$$

The second joint characteristic function reads

$$\Psi_Y(u_1, u_2) = \int_{-\infty}^{\infty} f_G(g)\left(\int_{-\infty}^{\infty} \lambda(\rho)\{\exp[\mathrm{j}u_1 h(t_1-\rho) + \mathrm{j}u_2 h(t_2-\rho)] - 1\}\,\mathrm{d}\rho\right)\mathrm{d}g \tag{8.67}$$

where $f_G(g)$ is the probability density function of G. Also in this case the process $Y(t)$ appears to be wide-sense stationary and the autocorrelation function becomes

$$R_{YY}(\tau) = \mathrm{E}[G^2]\ \mathrm{E}[\lambda]\ h(\tau) * h(-\tau) + \mathrm{E}^2[G]\ R_{\lambda\lambda}(\tau) * h(\tau) * h(-\tau) \tag{8.68}$$

and the power spectral density

$$\begin{aligned} S_{YY}(\omega) &= |H(\omega)|^2\ \{\mathrm{E}[G^2]\ \mathrm{E}[\lambda] + \mathrm{E}^2[G]\ S_{\lambda\lambda}(\omega)\} \\ &= |H(\omega)|^2\ \mathrm{E}^2[G]\ \left\{\frac{\mathrm{E}[G^2]}{\mathrm{E}^2[G]}\mathrm{E}[\lambda] + S_{\lambda\lambda}(\omega)\right\} \end{aligned} \tag{8.69}$$

The last term will in general comprise the information, whereas the first term is the shot noise spectral density. In applications where the random impulse amplitude G is an amplification generated by an optical or electronic device, the factor $\mathrm{E}[G^2]/\mathrm{E}^2[G]$ is called the excess noise factor. It is the factor by which the signal-to-shot noise ratio is decreased compared to the situation where such amplification is absent, i.e. G is constant and equal to 1. Since

$$\frac{\mathrm{E}[G^2]}{\mathrm{E}^2[G]} = \frac{\mathrm{E}^2[G] + \sigma_G^2}{\mathrm{E}^2[G]} = 1 + \frac{\sigma_G^2}{\mathrm{E}^2[G]} > 1 \tag{8.70}$$

the excess noise factor is always larger than 1.

Example 8.5:

A random-pulse process with the Poisson parameter $\lambda =$ constant is applied to a filter with the impulse response given by Equation (8.42). The G's are, independently from each other, selected from the set $G \in \{1, -1\}$ with equal probabilities. From these data and Campbell's theorem (Equations (8.65) and (8.66)), it follows that

$$\begin{aligned} \mathrm{E}[G] &= \mathrm{E}[Y] = 0 \\ \mathrm{E}[G^2] &= 1 \\ \sigma_Y^2 &= \lambda T \end{aligned} \tag{8.71}$$

This reduces Equation (8.69) to just the first term. The power spectrum of the output becomes

$$S_{YY}(\omega) = 4\lambda \frac{\sin^2\left(\frac{\omega T}{2}\right)}{\omega^2} \tag{8.72}$$

□

There are two main application areas for the processes dealt with in this section. The first example is the current produced by a photomultiplier tube or avalanche photodiode when an optical signal is detected. Each detected photon produces many electrons as a consequence of the internal amplification in the detector device. This process can be modelled as a Poisson impulse process, where G is a random variable with discrete integer amplitude. Looking at Equation (8.70) one may wonder why such amplification is applied when the signal-to-noise ratio is decreased by it. From the first line in this equation it is revealed that the information signal is amplified by $\mathrm{E}^2[G]$ and that is useful, since it raises the signal level with respect to the thermal noise (not taken into account in this chapter), which is dominant over the shot noise in the case of weak signal reception.

Secondly, when a radar on board an aircraft flying over the sea transmits pulses, it receives many copies of the transmitted pulse, called clutter. These echoes are randomly distributed in time and amplitude, due to reflections from the moving water waves. Similar reflections are received when flying over land, due to the relative changes in the earth's surface and eventual buildings. This process may be modelled as a random-pulse process.

8.6 SUMMARY

The Poisson distribution is recalled. Subsequently, the characteristic function is defined. This function provides a powerful tool for calculating moments of random variables. When taking the logarithm of the characteristic function, the second characteristic function results and it can be of help calculating moments of Poisson distributions and processes. Based on these functions several properties of the Poisson distribution are derived. The probability density function of the interarrival time is calculated.

The homogeneous Poisson process consists of a sequence of unit impulses with random distribution on the time axis and a Poisson distribution of the number of impulses per unit of time. When filtering this process a noise-like signal results, called shot noise. Based on the characteristic function and the second characteristic function several properties of this process are derived, such as the mean value, variance (Campbell's theorem) and autocorrelation function. From these calculations it follows that the homogeneous Poisson process is a wide-sense stationary process. Fourier transforming the autocorrelation function provides the power spectral density. Although part of the shot noise process behaves like white noise, it is not additive but multiplicative. Derivation of the properties of the inhomogeneous Poisson process and the random-pulse process is similar to that of the homogeneous Poisson process.

Although our approach emphasizes using the characteristic function to calculate the properties of random point processes, it should be stressed that this is not the only way to arrive at these results. A few application areas of random point processes are mentioned.

8.7 PROBLEMS

8.1 Consider two independent random variables X and Y. The random variable Z is defined as the sum $Z = X + Y$. Based on the use of characteristic functions show that the probability density function of Z equals the convolution of the probability density functions of X and Y.

8.2 Calculate the characteristic function of a Gaussian random variable.

8.3 Consider two jointly Gaussian random variables, X and Y, both having zero mean and variance of unity. Using the characteristic functions show that the probability density function of their sum, $Z = X + Y$, is Gaussian as well.

8.4 Use the characteristic function to calculate the variance of the waiting time of a Poisson process.

8.5 The characteristic function can be used to calculate the probability density function of a function $g(X)$ of a random variable if the transformation $Y = g(X)$ is one-to-one. Consider $Y = \sin X$, where X is uniformly distributed on the interval $(-\pi/2, \pi/2]$.

(a) Calculate the probability density function of Y.

(b) Use the result of (a) to calculate the probability density function of a full sine wave, i.e. $Y = \sin X$, where now X is uniformly distributed on the interval $(-\pi, \pi]$ *Hint:* extend the sine wave so that X uniformly covers the interval $(-\pi, \pi]$.

8.6 A circuit comprises 100 components. The circuit fails if one of the components fails. The time to failure of one component is exponentially distributed with a mean time to failure of 10 years. This distribution is the same for all components and the failure of each components is independent of the others.

(a) What is the probability that the circuit will be in operation for at least one year without interruption due to failure?

(b) What should the mean time to failure of a single component be so that the probability that the circuit will be in operation for at least one year without interruption is 0.9?

8.7 A switching centre has 100 incoming lines and one outgoing line. The arrival of calls is Poisson distributed with an average rate of 5 calls per hour. Suppose that each call lasts exactly 3 minutes.

(a) Calculate the probability that an incoming call finds the outgoing line blocked.

(b) The subscribers have a contract that guarantees a blocking probability less than 0.01. They complain that the blocking probability is higher than what was promised in the contract with the provider. Are they right?

(c) How many outgoing lines are needed to meet this condition in the contract?

8.8 Visitors enter a museum according to a Poisson distribution with a mean of 10 visitors per hour. Each visitor stays in the museum for exactly 0.5 hour.

(a) What is the mean value of the number of visitors present in the museum?

(b) What is the variance of the number of visitors present in the museum?

(c) What is the probability that there are no visitors in the museum?

8.9 Consider a shot noise process with constant parameter λ. This process is applied to a filter that has the impulse response $h(t) = \exp(-\alpha t)\,\mathrm{u}(t)$, with $\mathrm{u}(t)$ the unit step function.

(a) Find the mean value of the filtered shot noise process.

(b) Find the variance of the filtered process.

(c) Find the autocorrelation function of the filtered process.

(d) Calculate the power spectral density of the filtered process.

8.10 We want to consider the properties of a filtered shot noise process $Y(t)$ with constant parameter $\lambda \gg 1$. The problem is that when $\lambda \to \infty$ both the mean and the variance become infinitely large. Therefore we consider the normalized process with zero mean and variance of unity for all λ values:

$$\Xi(t) \triangleq \frac{Y(t) - \mathrm{E}[Y(t)]}{\sigma_Y} = \frac{Y(t) - A\lambda}{B\sqrt{\lambda}}$$

where

$$A \triangleq \int_{-\infty}^{\infty} h(\tau)\,\mathrm{d}\tau$$

$$B^2 \triangleq \int_{-\infty}^{\infty} h^2(\tau)\,\mathrm{d}\tau$$

Apply Equation (8.38) and the series expansion of Equation (8.39) to the process $\Xi(t)$ to prove that the characteristic function of $\Xi(t)$ tends to that of a Gaussian random variable (see the outcome of Problem 8.2) and thus the filtered Poisson process approaches a Gaussian process when the Poisson parameter λ becomes large.

8.11 The amplitudes of the impulses of a Poisson impulse process are independent and identically distributed with $\mathrm{P}(G_i = 1) = 0.6$ and $\mathrm{P}(G_i = 2) = 0.4$. The process is applied to a linear time-invariant filter with a rectangular impulse response of height 10 and duration T. The Poisson parameter λ is constant.

(a) Find the mean value of the filter output process.

(b) Find the autocorrelation function of the filter output process.

(c) Calculate the power spectral density of the output process.

8.12 The generation of hole–electron pairs in a photodiode is modelled as a Poisson process due to the random arrival of photons. When the optical wave is modulated by a randomly phased harmonic signal this arrival has a time-dependent rate of

$$\lambda = \lambda_0[1 + \cos(\omega_0 t - \Theta)]$$

with Θ a random variable that is uniformly distributed on the interval $(0, 2\pi]$, and where the cosine term is the information-carrying signal. Each hole–electron pair creates an impulse of height e, being the electron charge. The photodetector has a random internal gain that has only integer values and that is uniformly distributed on the interval $[0, 10]$. The travel time T in the detector is modelled as an impulse response

$$h(t) = \begin{cases} 1, & 0 \leq t < T \\ 0, & \text{elsewhere.} \end{cases}$$

(a) Calculate the mean value of the photodiode current and the power of the detected signal current.

(b) Calculate the shot noise variance.

(c) Calculate the the signal-to-noise ratio in dB for $\lambda_0 = 7.5 \times 10^{12}$, $T = 10^{-9}$ and $\omega_0 = 2\pi \times 0.5 \times 10^9$.

References

1. A. Papoulis and S.U Pillai, *Probability, Random Variables, and Stochastic Processes*, fourth edition, McGraw-Hill, 2002.
2. P.Z. Peebles, *Probability, Random Variables, and Random Signal Principles*, fourth edition, McGraw-Hill, 2001.
3. C.W. Helstrom, *Probability and Stochastic Processes for Engineers*, second edition, Macmillan, 1991.
4. W.B. Davenport Jr and J.L. Root, *An Introduction to the Theory of Random Signals and Noise*, McGraw-Hill, 1958; reissue by IEEE Press, 1987.
5. K.S. Shanmugam and A.M. Breipohl, *Random Signals: Detection, Estimation and Data Analysis*, Wiley, 1988.
6. L.W. Couch II, *Digital and Analog Communication Systems*, sixth edition, Prentice-Hall, 2001.
7. A. Papoulis, *The Fourier Integral and Its Applications*, McGraw-Hill, 1962.
8. W. van Etten and J. van der Plaats, *Fundamentals of Optical Fiber Communications*, Prentice-Hall, 1991.
9. J. Proakis and M. Salehi, *Communication Systems Engineering*, second edition, Prentice-Hall, 2002.
10. H. Baher, *Analog and Digital Signal Processing*, second edition, Wiley, 2001.
11. S. Haykin, *Communication Systems*, fourth edition, Wiley, 2001.
12. A. Papoulis, *Signal Analysis*, McGraw-Hill, 1977.
13. A.B. Carlson, P.B. Crilly and J.C. Rutledge, *Communication Systems*, fourth edition, McGraw-Hill, 2002.
14. J.M. Wozencraft and I.M. Jacobs, *Principles of Communication Engineering*, Wiley, 1965.
15. C.W. Helstrom, *Statistical Theory of Signal Detection*, second edition, Pergamon, 1968.
16. L.A. Pipes, *Applied Mathematics for Engineers and Physicists*, second edition, McGraw-Hill, 1958.
17. D.L. Snyder and M.I. Miller, *Random Point Processes in Time and Space*, second edition, Springer, 1991.

Introduction to Random Signals and Noise W. van Etten

Further Reading

Ash, C., *The Probability Tutoring Book*, IEEE Press, 1993.

Bracewell, R., *The Fourier Transform and Its Applications*, second edition, Cambridge University Press, 1978.

Childers, D.G., *Probability and Random Processes*, Irwin, 1997.

Davenport Jr, W.B., *Probability and Random Processes*, McGraw-Hill, 1970.

Feller, W., *An Introduction to Probability Theory and Its Applications*, Vol. I, third edition, Wiley, 1968.

Franks, L.E., *Signal Theory*, Prentice-Hall, 1969.

Gradshteyn, I.S. and Ryzhik, I.M., *Table of Integrals, Series and Products*, fourth edition, Academic Press, 1980.

Leon-Garcia, A., *Probability and Random Processes for Electrical Engineers*, second edition, Addison-Wesley, 1994.

McDonough, R.N. and Whalen, A.D., *Detection of Signals in Noise*, second edition, Academic Press, 1995.

Melsa, J.L. and Cohn, D.L., *Decision and Estimation Theory*, McGraw-Hill, 1978.

Middleton, D., *An Introduction to Statistical Communication Theory*, McGraw-Hill, 1960; reissue by IEEE Press, 1997.

Poor, H.V., *An Introduction to Signal Detection and Estimation*, second edition, Springer-Verlag, 1994.

Proakis, J., *Digital Communications*, fourth edition, McGraw-Hill, 2003.

Schwartz, M., *Information Transmission, Modulation, and Noise*, fourth edition, McGraw-Hill, 1990.

Schwartz, M. and Shaw, L., *Signal Processing: Discrete Spectral Analysis, Detection and Estimation*, McGraw-Hill, 1975.

Thomas, J.B., *An Introduction to Statistical Communication Theory*, Wiley, 1969.

Thomas, J.B., *An Introduction to Applied Probability and Random Processes*, Krieger, 1981.

Van Trees, H.L., *Detection, Estimation, and Modulation Theory*, Part I, Wiley, 1968.

Van Trees, H.L., *Detection, Estimation, and Modulation Theory*, Part II, Wiley, 1971.

Van Trees, H.L., *Detection, Estimation, and Modulation Theory*, Part III, Wiley, 1971.

Ziemer, R.E. and Tranter, W.H., *Principles of Communications: Systems, Modulation, and Noise*, fifth edition, Wiley, 2002.

Introduction to Random Signals and Noise W. van Etten

Appendix A
Representation of Signals in a Signal Space

A.1 LINEAR VECTOR SPACES

In order to facilitate the geometrical representation of signals, we treat them as vectors. Indeed, signals can be considered to behave like vectors, as will be shown in the sequel. For that purpose we recall the properties of linear vector spaces. A vector space is called a linear vector space if it satisfies the following conditions:

1. $$\mathbf{x} + \mathbf{y} = \mathbf{y} + \mathbf{x} \tag{A.1}$$
2. $$\mathbf{x} + (\mathbf{y} + \mathbf{z}) = (\mathbf{x} + \mathbf{y}) + \mathbf{z} \tag{A.2}$$
3. $$\alpha(\mathbf{x} + \mathbf{y}) = \alpha\mathbf{x} + \alpha\mathbf{y} \tag{A.3}$$
4. $$(\alpha + \beta)\mathbf{x} = \alpha\mathbf{x} + \beta\mathbf{x} \tag{A.4}$$

where $\mathbf{x}$ and $\mathbf{y}$ are arbitrary vectors and α and β are scalars.

In an n-dimensional linear vector space we define a so-called inner product as

$$\mathbf{x} \cdot \mathbf{y} \triangleq \sum_{i=1}^{n} x_i y_i \tag{A.5}$$

where x_i and y_i are the elements of $\mathbf{x}$ and $\mathbf{y}$, respectively. Two vectors $\mathbf{x}$ and $\mathbf{y}$ are said to be orthogonal if $\mathbf{x} \cdot \mathbf{y} = 0$. The norm of a vector $\mathbf{x}$ is denoted by $\| \mathbf{x} \|$ and we define it by

$$\| \mathbf{x} \| \triangleq \sqrt{\mathbf{x} \cdot \mathbf{x}} = \sqrt{\sum_{i=1}^{n} x_i^2} \tag{A.6}$$

Introduction to Random Signals and Noise W. van Etten

This norm has the following properties:

5. $\| \mathbf{x} \| \geq 0$ (A.7)
6. $\| \mathbf{x} \| = 0 \Longleftrightarrow \mathbf{x} = \mathbf{0}$ (A.8)
7. $\| \mathbf{x} + \mathbf{y} \| \leq \| \mathbf{x} \| + \| \mathbf{y} \|$ (A.9)
8. $\| \alpha \mathbf{x} \| = |\alpha| \cdot \| \mathbf{x} \|$ (A.10)

In general, we can state that the norm of a vector represents the distance from an arbitrary point described by the vector to the origin, or alternatively it is interpreted as the length of the vector. From Equation (A.9) we can readily derive the Schwarz inequality

$$|\mathbf{x} \cdot \mathbf{y}| \leq \| \mathbf{x} \| \cdot \| \mathbf{y} \| \tag{A.11}$$

A.2 THE SIGNAL SPACE CONCEPT

In this section we consider signals defined on the time interval $[a, b]$. As in the case of vectors, we define the inner product of two signals $x(t)$ and $y(t)$, but now the definition reads

$$\langle x(t), y(t) \rangle \triangleq \int_a^b x(t) y(t) \, \mathrm{d}t \tag{A.12}$$

Note that using this definition, signals behave like vectors, i.e. they show the properties 1 to 8 as given in the preceding section. This is readily verified by considering the properties of integrals.

Let us consider a set of orthonormal signals $\{\phi_i(t)\}$, i.e. signals that satisfy the condition

$$\langle \phi_i(t), \phi_j(t) \rangle = \delta_{ij} \triangleq \begin{cases} 1, & i = j, \\ 0, & i \neq j \end{cases} \tag{A.13}$$

where δ_{ij} denotes the well-known Kronecker delta. When all signals of a specific class can exactly be described as a linear combination of the members of such a signal set, we call it a complete orthonormal signal set; here we take the class of square integrable signals. In this case each arbitrary signal of that class is written as

$$s(t) = \sum_{i=1}^{n} s_i \phi_i(t) \tag{A.14}$$

and the sequence $\{s_i\}$ can also be written as a vector $\mathbf{s}$, where the s_i are the elements of the signal vector $\mathbf{s}$. Figure A.1 shows a circuit that reconstructs $s(t)$ from the set $\{s_i\}$ and the orthonormal signal set $\{\phi_i(t)\}$; this circuit follows immediately from Equation (A.14).

No limitations are placed on the integer n; even an infinite number of elements is allowed. The elements s_i are found by

$$s_i = \langle s(t), \phi_i(t) \rangle = \int_a^b s(t) \phi_i(t) \, \mathrm{d}t \tag{A.15}$$

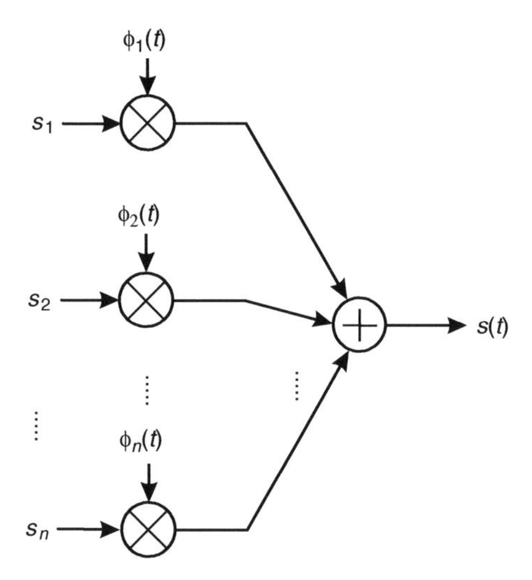

Figure A.1 A circuit that produces the signal $s(t)$ from its elements $\{s_i\}$ in the signal space

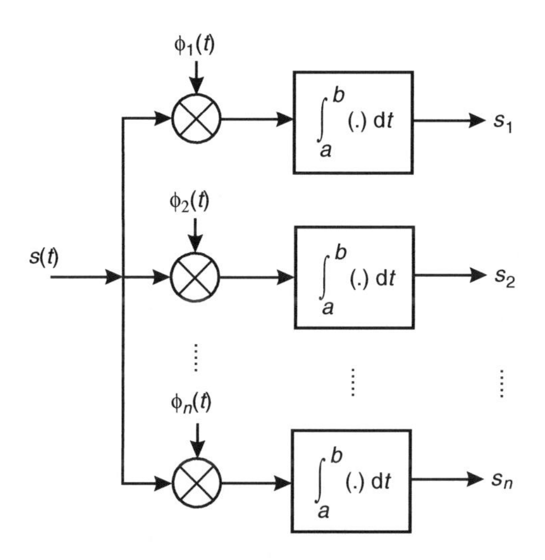

Figure A.2 A circuit that produces the signal space elements $\{s_i\}$ from the signal $s(t)$

Therefore, when we construct a vector space using the vector **s**, we have a geometrical representation of the signal $s(t)$. Along the ith axis we imagine the function $\phi_i(t)$; i.e. the set $\{\phi_i(t)\}$ is taken as the basis for the signal space. In fact, s_i indicates to what extent $\phi_i(t)$ contributes to $s(t)$. From Equation (A.15) a circuit is derived that produces the elements $\{s_i\}$ representing the signal $s(t)$ in the signal space $\{\phi_i(t)\}$; the circuit is given in Figure A.2.

The inner product of a signal vector with itself has an interesting interpretation:

$$\langle s(t), s(t) \rangle = \int_a^b s^2(t)\,\mathrm{d}t = \| \mathbf{s} \|^2 = \sum_i s_i^2 = E_s \tag{A.16}$$

with E_s the energy of the signal $s(t)$. From this equation it is concluded that the length of a vector in the signal space equals the square root of the signal energy. For purposes of detection it is important to establish that the distance between two signal vectors represents the square root of the energy of the difference of the two signals involved.

This concept of signal spaces is in fact a generalization of the well-known Fourier series expansion of signals.

Example A.1:

As an example let us consider the harmonic signal $s(t) = \mathrm{Re}\{a \exp[\mathrm{j}(\omega_0 t + \psi)]\}$, where $\mathrm{Re}\{\cdot\}$ is the real part of the expression in the braces. This signal is written as

$$s(t) = a \cos\psi \cos\omega_0 t - a \sin\psi \sin\omega_0 t \tag{A.17}$$

As the orthonormal signal set we consider $\{\sqrt{2}\cos(\omega_0 t)/\sqrt{T}, -\sqrt{2}\sin(\omega_0 t)/\sqrt{T}\}$ and the time interval $[a, b]$ is taken as $[0, T]$, with $T = k \times 2\pi/\omega_0$ and k integer. In this signal space, the signal $s(t)$ is represented by the vector $\mathbf{s} = [\sqrt{T}a \cos\psi/\sqrt{2}, \sqrt{T}a \sin\psi/\sqrt{2}]$. In fact, we have introduced in this way an alternative for the signal representation of harmonic signals in the complex plane. The elements of the vector **s** are recognized as the well-known quadrature I and Q signals.

□

A.3 GRAM–SCHMIDT ORTHOGONALIZATION

When we have an arbitrary set $\{f_i(t)\}$ of, let us say, N signals, then in general these signals will not be orthonormal. The Gram–Schmidt method shows us how to transform such a set into an orthonormal set, provided the members of the set $\{f_i(t)\}$ are linearly independent, i.e. none of the signal $f_i(t)$ can be written as a linear combination of the other signals. The first member of the orthonormal set is simply constructed as

$$\phi_1(t) \triangleq \frac{f_1(t)}{\sqrt{\langle f_1(t), f_1(t)\rangle}} \tag{A.18}$$

In fact, the signal $f_1(t)$ is normalized to the square root of its energy, or equivalently to the length of its corresponding vector in the signal space.

The second member of the orthonormal set is constructed by taking $f_2(t)$ and subtracting from this $f_2(t)$ the part that is already comprised of $\phi_1(t)$. In this way we arrive at the intermediate signal

$$g_2(t) \triangleq f_2(t) - \langle f_2(t), \phi_1(t)\rangle\phi_1(t) \tag{A.19}$$

Due to this operation the signal $g_2(t)$ will be orthogonal to $\phi_1(t)$. The functions $\phi_1(t)$ and $\phi_2(t)$ will become orthonormal if we construct $\phi_2(t)$ from $g_2(t)$ by normalizing it by its own length:

$$\phi_2(t) = \frac{g_2(t)}{\sqrt{\langle g_2(t), g_2(t)\rangle}} \tag{A.20}$$

Proceeding in this way we construct the kth intermediate signal by

$$g_k(t) \triangleq f_k(t) - \sum_{j=1}^{k-1} \langle f_k(t), \phi_j(t) \rangle \phi_j(t) \tag{A.21}$$

and by proper normalization we arrive at the kth member of the orthonormal signal set

$$\phi_k(t) \triangleq \frac{g_k(t)}{\sqrt{\langle g_k(t), g_k(t) \rangle}} \tag{A.22}$$

This procedure is continued until N orthonormal signals have been constructed. In case there are linear dependences, the dimensionality of the orthonormal signal space will be lower than N.

The orthonormal space that results from the Gram–Schmidt procedure is not unique; it will depend on the order in which the above described procedure is executed. Nevertheless, the geometrical signal constellation will not alter and the lengths of the vectors are invariant to the order chosen.

Example A.2:

Consider the signal set $\{f_i(t)\}$ given in Figure A.3(a). Since the norm of $f_1(t)$ is unity, we conclude that $\phi_1(t) = f_1(t)$. Moreover, from the figure it is deduced that the inner product $\langle f_2(t), \phi_1(t) \rangle = 0$ and $\langle f_2(t), f_2(t) \rangle = 4$, which means that $\phi_2(t) = f_2(t)/2$. Once this is known, it is easily verified that $\langle f_3(t), \phi_1(t) \rangle = \frac{1}{2}$ and $\langle f_3(t), \phi_2(t) \rangle = 0$, from which it follows that $g_3(t) = f_3(t) - \phi_1(t)/2$. The set of functions $\{g_i(t)\}$ has been depicted in Figure A.3(b). From this figure we calculate $\| g_3 \|^2 = \frac{1}{4}$, so that $\phi_3(t) = 2g_3(t)$, and finally from all those results the signal set $\{\phi_i(t)\}$, as given in Figure A.3(c), can be constructed. In this example the given functions do not show linear dependence and therefore the set $\{\phi_i(t)\}$ contains as many functions as the given set $\{f_i(t)\}$.

□

A.4 THE REPRESENTATION OF NOISE IN SIGNAL SPACE

In this section we will confine our analysis to the widely used concept of wide-sense stationary, zero mean, white, Gaussian noise. A sample function of the noise is denoted by $N(t)$ and the spectral density is $N_0/2$. We construct the noise vector $\mathbf{n}$ in the signal space, where the elements of this noise vector are defined by

$$n_i = \langle N(t), \phi_i(t) \rangle \tag{A.23}$$

Since this integration is a linear operation, these noise elements will also show a Gaussian distribution and it will be clear that the mean of n_i equals zero, for all i. When, besides these data, the cross-correlations of the different noise elements are determined, the noise

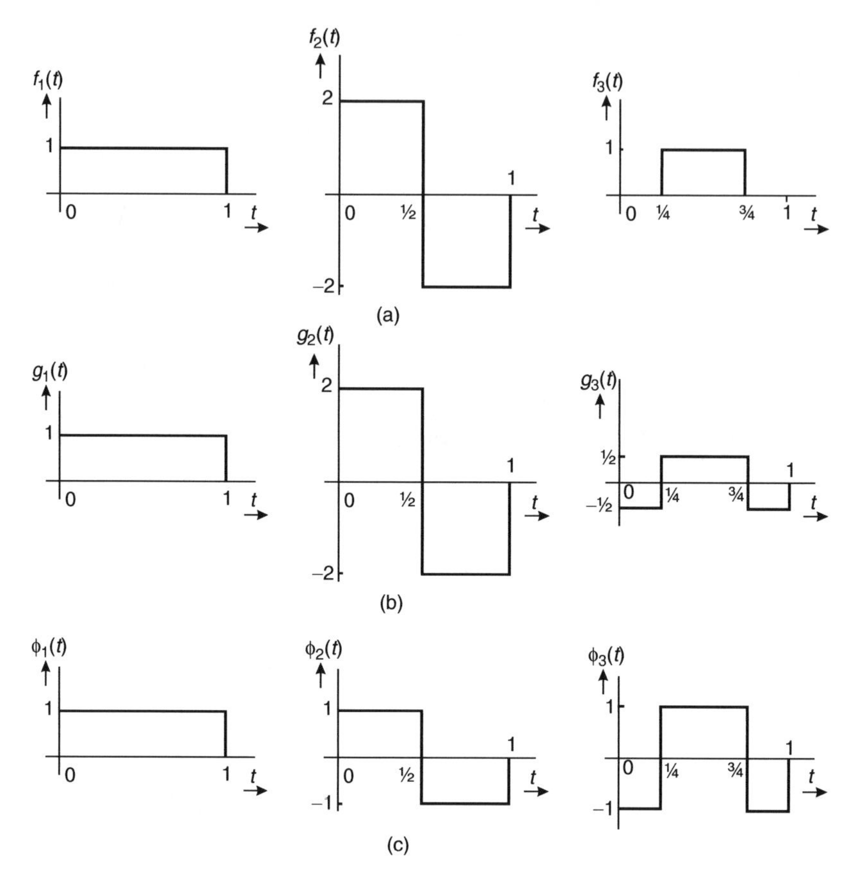

Figure A.3 The construction of a set of orthonormal functions $\{\phi_i(t)\}$ from a set of given functions $\{f_i(t)\}$ using the Gram–Schmidt orthogonalization procedure

elements are completely specified. Those cross-correlations are found to be

$$\begin{aligned}
\mathrm{E}[n_i n_j] &= \mathrm{E}\left[\int_a^b N(t)\ \phi_i(t)\,\mathrm{d}t \int_a^b N(\tau)\ \phi_j(\tau)\,\mathrm{d}\tau\right] \\
&= \mathrm{E}\left[\iint_a^b N(t)\ N(\tau)\ \phi_i(t)\ \phi_j(\tau)\,\mathrm{d}t\,\mathrm{d}\tau\right] \\
&= \iint_a^b \mathrm{E}[N(t)\,N(\tau)]\phi_i(t)\phi_j(\tau)\,\mathrm{d}t\,\mathrm{d}\tau
\end{aligned} \tag{A.24}$$

In this expression the expectation represents the autocorrelation function of the noise, which reads $R_{NN}(t,\tau) = \delta(t-\tau)N_0/2$. This is inserted in the last equation to arrive at

$$\begin{aligned}
\mathrm{E}[n_i n_j] &= \frac{N_0}{2}\iint_a^b \delta(t-\tau)\phi_i(t)\phi_j(\tau)\,\mathrm{d}t\,\mathrm{d}\tau \\
&= \frac{N_0}{2}\int_a^b \phi_i(t)\phi_j(t)\,\mathrm{d}t
\end{aligned} \tag{A.25}$$

Remembering that the set $\{\phi_i(t)\}$ is orthonormal, the following cross-correlations result

$$\mathrm{E}[n_i n_j] = \begin{cases} \dfrac{N_0}{2}, & \text{for } i = j \\ 0, & \text{for } i \neq j \end{cases} \tag{A.26}$$

From this equation it is concluded that the different noise elements n_i are uncorrelated and, since they are Gaussian with zero mean, they are independent. Moreover, all noise elements show the same variance of $N_0/2$. This simple and symmetric result is another interesting feature of the orthonormal signal spaces as introduced in this appendix.

A.4.1 Relevant and Irrelevant Noise

When considering a signal that is disturbed by noise we want to construct a common signal space to describe both the signal and noise in the same space. In that case we construct a signal space to completely describe all possible signals involved. When we want to attempt to describe the noise $N(t)$ using that signal space, it will, as a rule, be inadequate to completely characterize the noise. In that case we split the noise into one part $N_\mathrm{r}(t)$ that is projected on to the signal space, called the relevant noise, and another part $N_\mathrm{i}(t)$ that is orthogonal to the space set up by the signals, called the irrelevant noise. Thus, the relevant noise is given by the vector $\mathbf{n}_\mathrm{r}$ with components

$$n_{\mathrm{r},i} = \int_a^b N(t)\phi_i(t)\,\mathrm{d}t \tag{A.27}$$

By definition the irrelevant noise reads

$$N_\mathrm{i}(t) \triangleq N(t) - N_\mathrm{r}(t) \tag{A.28}$$

Next we will show that the irrelevant noise part is orthogonal to the signal space. For this purpose let us consider a signal $s(t)$ and an orthonormal signal space $\{\phi_i(t)\}$. Let us suppose that $s(t)$ can completely be described as a vector in this signal space. The inner product of the irrelevant noise $N_\mathrm{i}(t)$ and the signal reads

$$\begin{aligned} \int_a^b N_\mathrm{i}(t)s(t)\,\mathrm{d}t &= \int_a^b \{N(t) - N_\mathrm{r}(t)\}s(t)\,\mathrm{d}t \\ &= \int_a^b N(t)s(t)\,\mathrm{d}t - \int_a^b N_\mathrm{r}(t)s(t)\,\mathrm{d}t \\ &= \int_a^b N(t)\sum_k s_k\phi_k(t)\,\mathrm{d}t - \int_a^b \sum_k n_{\mathrm{r},k}\phi_k(t)s(t)\,\mathrm{d}t \\ &= \int_a^b \sum_k s_k N(t)\phi_k(t)\,\mathrm{d}t - \sum_k n_{\mathrm{r},k}s_k \\ &= \sum_k s_k n_{\mathrm{r},k} - \sum_k n_{\mathrm{r},k}s_k = 0 \end{aligned} \tag{A.29}$$

For certain applications, for instance optimum detection of a known signal in noise, it appears that the irrelevant part of the noise may be discarded.

A.5 SIGNAL CONSTELLATIONS

In this section we will present the signal space description of a few signals that are often met in practice. The signal space with indications of possible signal realizations is called a signal constellation. In detection, the error probability appears always to be a function of E_d/N_0, where E_d is the energy of the difference of signals. However, we learned in Section A.2 that this energy is in the signal space represented by the squared distance of the signals involved. Considering two signals $s_i(t)$ and $s_j(t)$, their squared distance is written as

$$d_{ij}^2 = \| \mathbf{s}_i - \mathbf{s}_j \|^2 = \int_a^b [s_i(t) - s_j(t)]^2 \, \mathrm{d}t = E_i + E_j - 2E_{ij} \tag{A.30}$$

where E_i and E_j are the energies of the corresponding signals and E_{ij} represents the inner product of the signals. In specific cases where $E_i = E_j = E$ for all i and j, Equation (A.30) is written as

$$d_{ij}^2 = 2E(1 - \rho_{ij}) \tag{A.31}$$

with the cross-correlations ρ_{ij} defined by

$$\rho_{ij} \triangleq \frac{\mathbf{s}_i \cdot \mathbf{s}_j}{\| \mathbf{s}_i \| \cdot \| \mathbf{s}_j \|} \tag{A.32}$$

A.5.1 Binary Antipodal Signals

Consider the two rectangular signals

$$s_1(t) = -s_2(t) = \sqrt{\frac{E}{T}}, \quad 0 \leq t \leq T \tag{A.33}$$

and their bandpass equivalents

$$s_1(t) = -s_2(t) = \sqrt{\frac{2E}{T}} \cos \omega_0 t, \quad 0 \leq t \leq T \tag{A.34}$$

with $T = n \times 2\pi/\omega_0$ and n integer.

The cross-correlation of those signals equals -1 and the distance between the two signals is

$$d_{12} = \sqrt{2E(1 - \rho_{12})} = 2\sqrt{E} \tag{A.35}$$

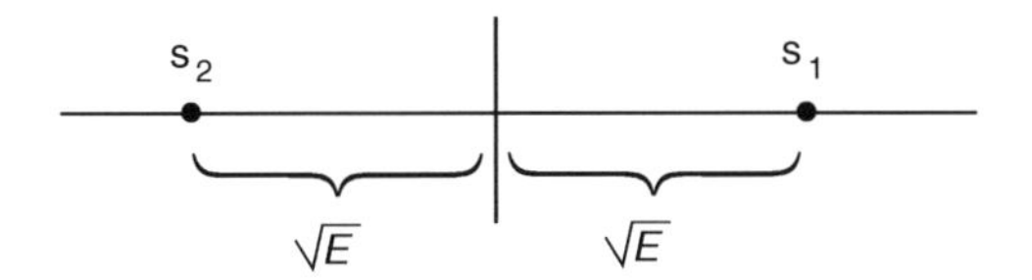

Figure A.4 The signal constellation of the binary antipodal signals

The space is one-dimensional, since the two signals involved are dependent. The signal vectors are $\mathbf{s}_1 = [\sqrt{E}]$ and $\mathbf{s}_2 = [-\sqrt{E}]$. This signal set is called an antipodal signal constellation and is depicted in Figure A.4.

A.5.2 Binary Orthogonal Signals

Next we consider the signal set

$$s_1(t) = \sqrt{\frac{2E}{T}} \cos \omega_0 t, \qquad 0 \leq t \leq T \tag{A.36}$$

$$s_2(t) = \sqrt{\frac{2E}{T}} \sin \omega_0 t, \qquad 0 \leq t \leq T \tag{A.37}$$

where either $T = n\pi/\omega_0$ (with n integer) or $T \gg 1/\omega_0$, so that $\rho_{12} = 0$ or $\rho_{12} \approx 0$, respectively. Due to this property those signals are called orthogonal signals. In this case the signal space is two-dimensional and $\mathbf{s}_1 = [\sqrt{E}, 0]$, while $\mathbf{s}_2 = [0, \sqrt{E}]$. It is easily verified that the distance between the signals amounts to $d_{12} = \sqrt{2E}$. The signal constellation is given in Figure A.5.

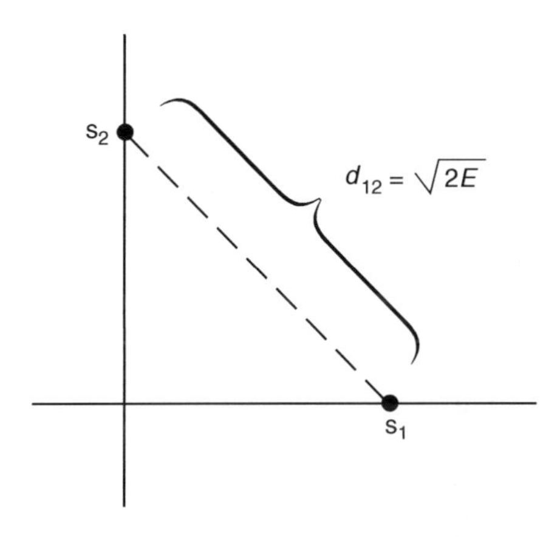

Figure A.5 The signal constellation of the binary orthogonal signals

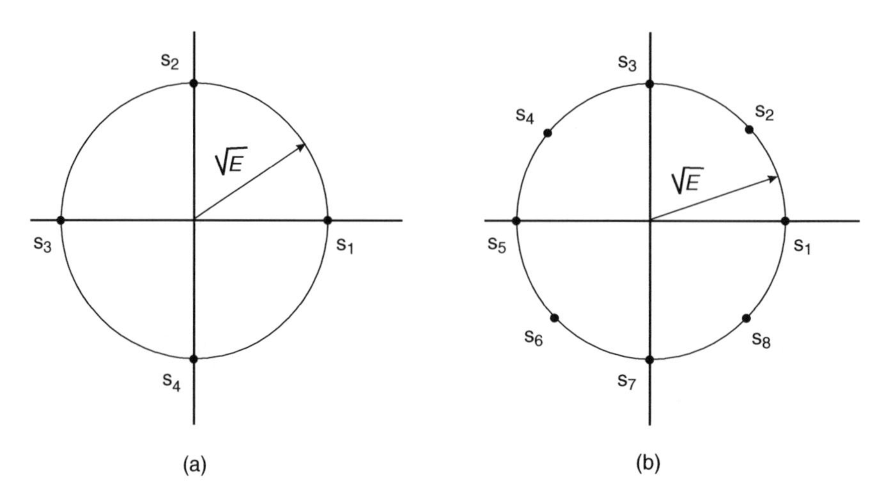

Figure A.6 The signal constellation of M-phase signals

A.5.3 Multiphase Signals

In the multiphase case all possible signal vectors are on a circle with radius $\sqrt{E}$. This corresponds to the M-ary phase modulation. The signals are represented by

$$s_i(t) = \mathrm{Re}\left\{\sqrt{\frac{2E}{T}}\exp\left(\mathrm{j}\omega_0 t + \mathrm{j}(i-1)\frac{2\pi}{M}\right)\right\}, \quad \text{for } i = 1, 2, \ldots, M;\ 0 \le t \le T \tag{A.38}$$

with the same requirement for the relation between T and ω_0 as in the former section. The signal vectors are given by

$$\mathbf{s}_i = \left[\sqrt{E}\cos\frac{(i-1)2\pi}{M}, \sqrt{E}\sin\frac{(i-1)2\pi}{M}\right] \tag{A.39}$$

In Figure A.6 two examples are depicted, namely $M = 4$ in Figure A.6(a) and $M = 8$ in Figure A.6(b). The case of $M = 4$ can be considered as a pair of two orthogonal signals; namely the pair of vectors $[\mathbf{s}_1, \mathbf{s}_3]$ is orthogonal to the pair $[\mathbf{s}_2, \mathbf{s}_4]$. For that reason this is a special case of the biorthogonal signal set, which is dealt with later on in this section. This orthogonality is used in QPSK modulation.

A.5.4 Multiamplitude Signals

The multiamplitude case is a straightforward extension of the antipodal signal constellation. The extension is in fact a manifold, let us say M, of signal vectors on the one-dimensional axis (see Figure A.4). In most applications these points are equidistant.

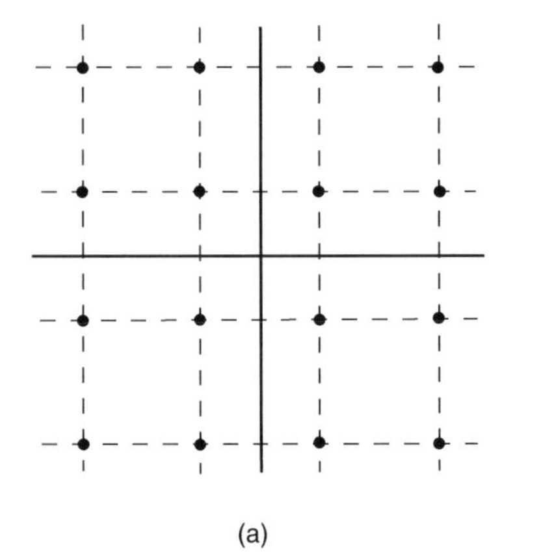

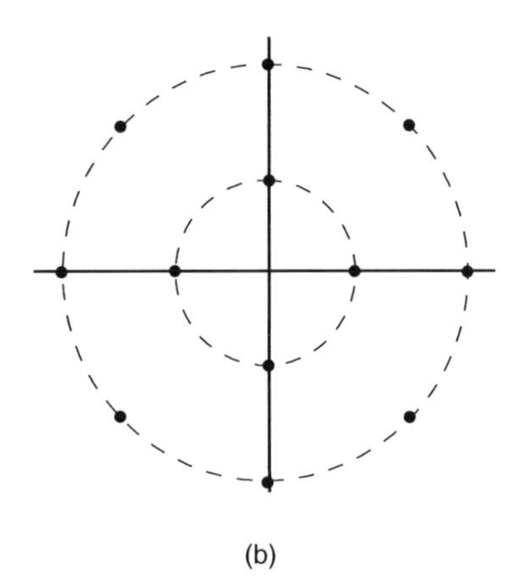

Figure A.7 The signal constellation of QAM signals: (a) rectangular distribution; (b) circular distribution

A.5.5 QAM Signals

A QAM (quadrature amplitude modulated) signal is a signal where both the amplitude and phase are modulated. In that sense it is a combination of the multiamplitude and multiphase modulated signal. Different constellations are possible; two of them are depicted in Figure A.7. Figure A.7(a) shows a rectangular grid of possible signal vectors, whereas in the example of Figure A.7(b) the vectors are situated on circles. This signal constellation is used in such applications as high-speed telephone modems, cable modems and digital distribution of audio and video signals over CATV networks.

A.5.6 M-ary Orthogonal Signals

The M-ary orthogonal signal set is no more no less than an M-dimensional extension of the binary orthogonal signal set; i.e. in this case the signal space has M dimensions and all possible signals are orthogonal to all others. For $M = 3$ and assuming that all signals bear the same energy, the signal vectors are

$$\begin{aligned} \mathbf{s}_1 &= [\sqrt{E}, 0, 0] \\ \mathbf{s}_2 &= [0, \sqrt{E}, 0] \\ \mathbf{s}_3 &= [0, 0, \sqrt{E}] \end{aligned} \tag{A.40}$$

This signal set has been illustrated in Figure A.8. The distance between two arbitrary signal pairs is $d = \sqrt{2E}$.

A.5.7 Biorthogonal Signals

An M-ary biorthogonal signal set is constructed from an $M/2$-ary orthogonal signal set by simply adding the negatives of all the orthogonal signals. It will be clear that the dimension

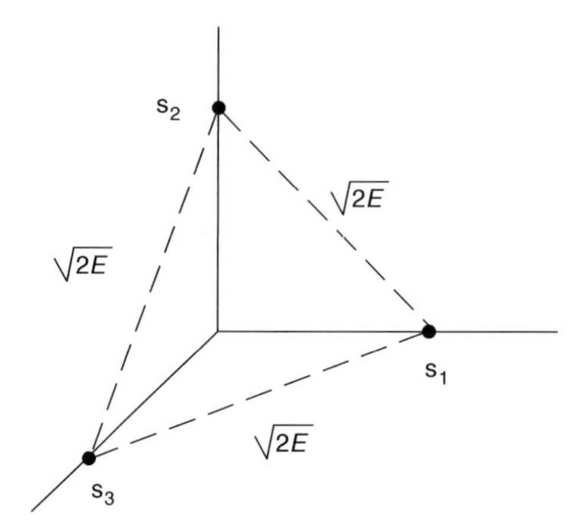

Figure A.8 The signal constellation of the M-ary orthogonal signal set for $M = 3$

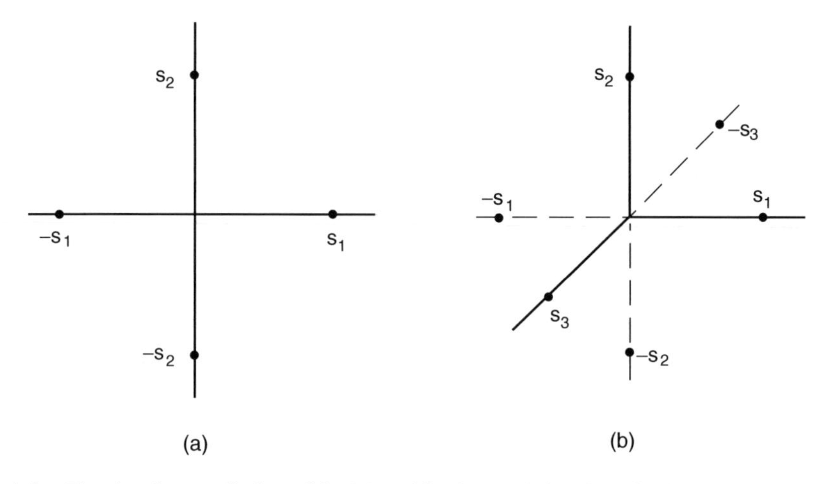

Figure A.9 The signal constellation of the M-ary biorthogonal signal set for (a) $M = 4$ and (b) $M = 6$

of the signal space remains as $M/2$. The result is given in Figure A.9(a) for $M = 4$ and in Figure A.9(b) for $M = 6$.

A.5.8 Simplex Signals

To explain the simplex signal set we start from a set of M orthogonal signals $\{f_i(t)\}$ with the vector presentation $\{\mathbf{f}_i\}$. We determine the mean of this signal set

$$\bar{\mathbf{f}} = \frac{1}{M}\sum_{i=1}^{M}\mathbf{f}_i \tag{A.41}$$

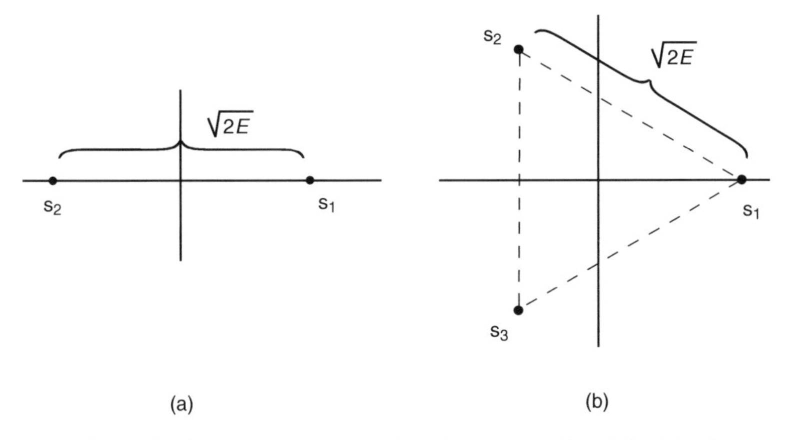

Figure A.10 The simplex signal set for (a) $M = 2$ and (b) $M = 3$

In order to arrive at the simplex signal set, this mean is subtracted from each vector of the orthogonal set:

$$\mathbf{s}_i = \mathbf{f}_i - \bar{\mathbf{f}}, \quad i = 1, 2, \ldots, M \tag{A.42}$$

In fact this operation means that the origin is translated to the point $\bar{\mathbf{f}}$. Therefore the distance between any pair of possible vectors $\mathbf{s}_i$ remains $d = \sqrt{2E}$. Figure A.10 shows the simplex signals for $M = 2$ and $M = 3$. Note that the dimensionality is reduced by 1 compared to the starting orthogonal set.

Due to the transformation the signals are no longer orthogonal. On the other hand, it will be clear that the mean signal energy has decreased. Since the distance between signal pairs are the same as for orthogonal signals, the simplex signal set is able to realize communication with the same quality (i.e. error probability) but using less energy per bit and thus less power.

A.6 PROBLEMS

A.1 Derive the Schwarz inequality (A.11) from the triangular inequality (A.9).

A.2 Use the properties of integration and the definition of Equation (A.12) to show that signals satisfy the properties of vectors as given in Section A.1.

A.3 Use Equation (A.19) to show that $g_2(t)$ is orthogonal to $\phi_1(t)$.

A.4 Show that for binary orthogonal signals one of the two given relations between T and ω_0 is required for orthogonality.

A.5 Calculate the distance between adjacent signal vectors, i.e. the minimum distance, for the M-ary phase signal.

A.6 Calculate the energy of the signals from the simplex signal set, expressed in terms of the energy in the signals of the M-ary orthogonal signal set.

A.7 Calculate the cross-correlation between the various signals of the simplex signal set.

Appendix B
Attenuation, Phase Shift and Decibels

The transfer function of a linear time-invariant system, denoted by $H(\omega)$, is in general a complex function of ω. It is indicative of the manner in which time harmonic signals (sine and cosine) propagate through such a system. The modulus of $H(\omega)$ is the factor by which the amplitude of the incoming harmonic signal is scaled and the argument of $H(\omega)$ indicates the phase shift introduced by the system. We then can write

$$H(\omega) = \exp[-a(\omega) - \mathrm{j}b(\omega)] = \exp[-a(\omega)]\,\exp[-\mathrm{j}b(\omega)] = |H(\omega)|\,\exp[-\mathrm{j}b(\omega)] \tag{B.1}$$

where

$$a(\omega) = \ln\frac{1}{|H(\omega)|} \tag{B.2}$$

is the attenuation of the system in neper (abbreviated as Np) and $b(\omega)$ is the phase shift introduced by the system. The neper is an old unit that is not used in practice anymore. A more common measure of attenuation is the decibel (abbreviated as dB). The attenuation in dB is defined as

$$a_{\mathrm{d}}(\omega) = 20\,\log_{10}\frac{1}{|H(\omega)|} \tag{B.3}$$

From the relation

$$\log_{10} x = \frac{\ln x}{\ln 10} \tag{B.4}$$

we conclude that

$$a_{\mathrm{d}}(\omega) = 20\frac{1}{\ln 10}\ln\frac{1}{|H(\omega)|} = 20 \times 0.4343\ln\frac{1}{|H(\omega)|} = 8.686\,a(\omega) \tag{B.5}$$

Introduction to Random Signals and Noise W. van Etten

From this it follows that 1 Np is equivalent to 8.686 dB. Because the neper is used infrequently, we will drop the subscript 'd' and henceforth speak about attenuation exclusively in terms of dB and denote this as a. The transfer function $H(\omega)$ is usually defined as a ratio of the output voltage (or current) and the input voltage (or current). If we look at the power ratio between the input and output, it is evident that the powers are proportional to the square of the voltages or currents, i.e.

$$\frac{P_\mathrm{i}}{P_\mathrm{o}} = \frac{|I_\mathrm{i}|^2 R_\mathrm{i}}{|I_\mathrm{o}|^2 R_\mathrm{o}} = \frac{\dfrac{|V_\mathrm{i}|^2}{R_\mathrm{i}}}{\dfrac{|V_\mathrm{o}|^2}{R_\mathrm{o}}} \tag{B.6}$$

where the indices 'o' and 'i' indicate that the quantity is related to the output and input, respectively. The quantities $R_{\mathrm{o,i}}$ give the real part of the impedances. If we assume the situation where $R_\mathrm{o} = R_\mathrm{i}$, which is often the case, then the power attenuation is found as

$$\begin{aligned} a &= 20 \times \log_{10} \frac{|V_\mathrm{i}|}{|V_\mathrm{o}|} = 20 \times \log_{10} \frac{|I_\mathrm{i}|}{|I_\mathrm{o}|} \\ &= 10 \times \log_{10} \frac{P_\mathrm{i}}{P_\mathrm{o}} \end{aligned} \tag{B.7}$$

The advantage in using a logarithmic measure for the ratio of magnitudes between the inputs and outputs lies in the fact that in series connections of systems, the attenuation in each of the systems expressed in dB needs only to be summed in order to obtain the total attenuation in dB.

Furthermore, we mention that the decibel is also used to express the absolute levels of power, current and voltage. The following notations are among others in use:

- dBW – power with respect to $P_0 = 1$ W
- dBm – power with respect to $P_0 = 1$ mW
- dBμV – voltage with respect to $V_0 = 1\ \mu$V.

Definitions and general properties of logarithms are given in Appendix C, Section C.6. Table B.1 presents a list of linear power ratios and the corresponding amount of dBs.

Table B.1 List of dB values for a given linear power ratio

Linear ratio	dB
0.25	−6
0.5	−3
1	0
2	3
4	6
10	10
100	20
1000	30

Appendix C
Mathematical Relations

C.1 TRIGONOMETRIC RELATIONS

$$\cos(\alpha \pm \beta) = \cos\alpha \cos\beta \mp \sin\alpha \sin\beta \tag{C.1}$$

$$\sin(\alpha \pm \beta) = \sin\alpha \cos\beta \pm \cos\alpha \sin\beta \tag{C.2}$$

$$\cos(\alpha \pm \frac{\pi}{2}) = \mp \sin\alpha \tag{C.3}$$

$$\sin(\alpha \pm \frac{\pi}{2}) = \pm \cos\alpha \tag{C.4}$$

$$\cos 2\alpha = \cos^2\alpha - \sin^2\alpha \tag{C.5}$$

$$\sin 2\alpha = 2\sin\alpha \cos\alpha \tag{C.6}$$

$$\cos\alpha = \frac{1}{2}[\exp(\mathrm{j}\alpha) + \exp(-\mathrm{j}\alpha)] \tag{C.7}$$

$$\sin\alpha = \frac{1}{2\mathrm{j}}[\exp(\mathrm{j}\alpha) - \exp(-\mathrm{j}\alpha)] \tag{C.8}$$

$$\cos\alpha \cos\beta = \frac{1}{2}[\cos(\alpha-\beta) + \cos(\alpha+\beta)] \tag{C.9}$$

$$\sin\alpha \sin\beta = \frac{1}{2}[\cos(\alpha-\beta) - \cos(\alpha+\beta)] \tag{C.10}$$

$$\sin\alpha \cos\beta = \frac{1}{2}[\sin(\alpha-\beta) + \sin(\alpha+\beta)] \tag{C.11}$$

$$\cos^2\alpha = \frac{1}{2}(1 + \cos 2\alpha) \tag{C.12}$$

$$\sin^2\alpha = \frac{1}{2}(1 - \cos 2\alpha) \tag{C.13}$$

$$\cos^3\alpha = \frac{3}{4}\cos\alpha + \frac{1}{4}\cos 3\alpha \tag{C.14}$$

$$\sin^3\alpha = \frac{3}{4}\sin\alpha - \frac{1}{4}\sin 3\alpha \tag{C.15}$$

Introduction to Random Signals and Noise W. van Etten

$$\cos^4\alpha = \frac{3}{8} + \frac{1}{2}\cos 2\alpha + \frac{1}{8}\cos 4\alpha \tag{C.16}$$

$$\sin^4\alpha = \frac{3}{8} - \frac{1}{2}\cos 2\alpha + \frac{1}{8}\cos 4\alpha \tag{C.17}$$

$$A\cos\alpha - B\sin\alpha = R\cos(\alpha + \theta) \tag{C.18}$$

where

$$R = \sqrt{A^2 + B^2}, \qquad A = R\cos\theta$$
$$\theta = \arctan(B/A), \quad B = R\sin\theta$$

C.2 DERIVATIVES

$y = y(x)$	$y' = \mathrm{d}y/\mathrm{d}x$	
$y = a$	$y' = 0, \quad a$ constant	(C.19)
$y = x$	$y' = 1$	(C.20)
$y = x^n$	$y' = nx^{n-1}$	(C.21)
$y = \ln x, \quad x > 0$	$y' = 1/x$	(C.22)
$y = \log x, \quad x > 0$	$y' = (1/x)\log \mathrm{e}$	(C.23)
$y = \sin x$	$y' = \cos x$	(C.24)
$y = \cos x$	$y' = -\sin x$	(C.25)
$y = \tan x$	$y' = 1/\cos^2 x$	(C.26)
$y = \cot x$	$y' = -1/\sin^2 x$	(C.27)
$y = \exp x$	$y' = \exp x$	(C.28)
$y = a^x$	$y' = a^x \ln a$	(C.29)
$y = \arcsin x, \quad \lvert x\rvert < 1$	$y' = 1/\sqrt{1 - x^2}$	(C.30)
$y = \arccos x, \quad \lvert x\rvert < 1$	$y' = -1/\sqrt{1 - x^2}$	(C.31)
$y = \arctan x$	$y' = 1/(1 + x^2)$	(C.32)
$y = \sin(ax + b)$	$y' = a\cos(ax + b)$	(C.33)

C.2.1 Rules for Differentiation

$y = u + v$	$y' = u' + v'$	(C.34)
$y = uv$	$y' = u'v + uv'$	(C.35)
$y = u\,v\,w$	$y' = u'vw + uv'w + uvw'$	(C.36)
$y = u/v$	$y' = (u'v - uv')/v^2$	(C.37)
$y = u^v$	$y' = u^v(v'\ln u + u'v/u)$	(C.38)

C.2.2 Chain Rule

$$y = f(z), \qquad z = g(x) \rightarrow \qquad y' = \frac{\mathrm{d}y}{\mathrm{d}z} \times \frac{\mathrm{d}z}{\mathrm{d}x} \tag{C.39}$$

C.2.3 Stationary Points

$$\text{maximum if: } \quad f'(x) = 0 \text{ and } f''(x) < 0 \tag{C.40}$$

$$\text{minimum if: } \quad f'(x) = 0 \text{ and } f''(x) > 0 \tag{C.41}$$

$$\text{point of inflection if: } \quad f''(x) = 0 \text{ and } f'''(x) \neq 0 \tag{C.42}$$

C.3 INDEFINITE INTEGRALS

C.3.1 Basic Integrals

$$\int f'(x)\,\mathrm{d}x = f(x) \tag{C.43}$$

$$\int_a^b f(x)\,\mathrm{d}x = F(b) - F(a), \quad \text{if } \int f(x)\,\mathrm{d}x = F(x) \tag{C.44}$$

$$\int x^n\,\mathrm{d}x = \frac{x^{n+1}}{n+1} \tag{C.45}$$

$$\int \frac{1}{x}\,\mathrm{d}x = \ln|x| \tag{C.46}$$

$$\int \sin x\,\mathrm{d}x = -\cos x \tag{C.47}$$

$$\int \cos x\,\mathrm{d}x = \sin x \tag{C.48}$$

$$\int \tan x\,\mathrm{d}x = -\ln\cos x \tag{C.49}$$

$$\int \frac{1}{\sqrt{1-x^2}}\,\mathrm{d}x = \arcsin x = \frac{\pi}{2} - \arccos x, \quad |x| < 1 \tag{C.50}$$

$$\int \frac{1}{1+x^2}\,\mathrm{d}x = \arctan x = \frac{\pi}{2} - \operatorname{arccot} x \tag{C.51}$$

$$\int \exp x\,\mathrm{d}x = \exp x \tag{C.52}$$

$$\int a^x\,\mathrm{d}x = \frac{a^x}{\ln a}, \quad a > 0 \tag{C.53}$$

$$\int \sinh x\,\mathrm{d}x = \cosh x \tag{C.54}$$

$$\int \cosh x\,\mathrm{d}x = \sinh x \tag{C.55}$$

$$\int \frac{1}{\sin x}\,\mathrm{d}x = \ln\left(\tan\frac{x}{2}\right) \tag{C.56}$$

$$\int \frac{1}{\cos^2 x}\,\mathrm{d}x = \tan x \tag{C.57}$$

$$\int \frac{1}{\sin^2 x}\,\mathrm{d}x = -\cot x \tag{C.58}$$

C.3.2 Integration by Parts

$$\int u\,\mathrm{d}v = u\,v - \int v\,\mathrm{d}u \quad \text{or} \quad \int u\,v'\,\mathrm{d}x = u\,v - \int u'\,v\,\mathrm{d}x \tag{C.59}$$

$$\text{example :} \quad \int \ln x\,\mathrm{d}x = x\ln x - \int x\,\frac{1}{x}\,\mathrm{d}x = x(\ln x - 1)$$

C.3.3 Rational Algebraic Functions

$$\int (a+bx)^n\,\mathrm{d}x = \frac{(a+bx)^{n+1}}{b(n+1)}, \quad n > 0 \tag{C.60}$$

$$\int \frac{1}{a+bx}\,\mathrm{d}x = \frac{1}{b}\ln|a+bx| \tag{C.61}$$

$$\int \frac{1}{(a+bx)^n}\,\mathrm{d}x = \frac{-1}{b(n-1)(a+bx)^{n-1}}, \quad n > 1 \tag{C.62}$$

$$\begin{aligned}\int \frac{1}{ax^2+bx+c}\,\mathrm{d}x &= \frac{2}{\sqrt{4ac-b^2}}\arctan\left(\frac{2ax+b}{\sqrt{4ac-b^2}}\right), && \text{for } b^2 < 4ac\\ &= \frac{1}{\sqrt{b^2-4ac}}\ln\left|\frac{2ax+b-\sqrt{b^2-4ac}}{2ax+b+\sqrt{b^2-4ac}}\right|, && \text{for } b^2 > 4ac\\ &= \frac{-2}{2ax+b}, && \text{for } b^2 = 4ac\end{aligned} \tag{C.63}$$

$$\int \frac{x}{ax^2+bx+c}\,\mathrm{d}x = \frac{1}{2a}\ln|ax^2+bx+c| - \frac{b}{2a}\int \frac{1}{ax^2+bx+c}\,\mathrm{d}x \tag{C.64}$$

$$\int \frac{1}{a^2+b^2x^2}\,\mathrm{d}x = \frac{1}{ab}\arctan\left(\frac{bx}{a}\right) \tag{C.65}$$

$$\int \frac{x}{a^2+x^2}\,\mathrm{d}x = \frac{1}{2}\ln(a^2+x^2) \tag{C.66}$$

$$\int \frac{x^2}{a^2+b^2x^2}\,\mathrm{d}x = \frac{x}{b^2} - \frac{a}{b^3}\arctan\left(\frac{bx}{a}\right) \tag{C.67}$$

$$\int \frac{1}{(a^2+x^2)^2}\,\mathrm{d}x = \frac{x}{2a^2(a^2+x^2)} + \frac{1}{2a^3}\arctan\left(\frac{x}{a}\right) \tag{C.68}$$

$$\int \frac{x}{(a^2+x^2)^2}\,\mathrm{d}x = \frac{-1}{2(a^2+x^2)} \tag{C.69}$$

$$\int \frac{x^2}{(a^2+x^2)^2}\,\mathrm{d}x = \frac{-x}{2(a^2+x^2)} + \frac{1}{2a}\arctan\left(\frac{x}{a}\right) \tag{C.70}$$

$$\int \frac{1}{(a^2+x^2)^3}\,\mathrm{d}x = \frac{x}{4a^2(a^2+x^2)^2} + \frac{3x}{8a^4(a^2+x^2)} + \frac{3}{8a^5}\arctan\left(\frac{x}{a}\right) \tag{C.71}$$

$$\int \frac{x^2}{(a^2+x^2)^3}\,\mathrm{d}x = \frac{-x}{4(a^2+x^2)^2} + \frac{x}{8a^2(a^2+x^2)} + \frac{1}{8a^3}\arctan\left(\frac{x}{a}\right) \tag{C.72}$$

$$\int \frac{x^4}{(a^2+x^2)^3}\,\mathrm{d}x = \frac{a^2 x}{4(a^2+x^2)^2} - \frac{5x}{8(a^2+x^2)} + \frac{3}{8a}\arctan\left(\frac{x}{a}\right) \tag{C.73}$$

$$\int \frac{1}{(a^2+x^2)^4}\,\mathrm{d}x = \frac{x}{6a^2(a^2+x^2)^3} + \frac{5x}{24a^4(a^2+x^2)^2} + \frac{5x}{16a^6(a^2+x^2)} + \frac{5}{16a^7}\arctan\left(\frac{x}{a}\right) \tag{C.74}$$

$$\int \frac{x^2}{(a^2+x^2)^4}\,\mathrm{d}x = \frac{-x}{6(a^2+x^2)^3} + \frac{x}{24a^2(a^2+x^2)^2} + \frac{x}{16a^4(a^2+x^2)} + \frac{1}{16a^5}\arctan\left(\frac{x}{a}\right) \tag{C.75}$$

$$\int \frac{x^4}{(a^2+x^2)^4}\,\mathrm{d}x = \frac{a^2 x}{6(a^2+x^2)^3} - \frac{7x}{24(a^2+x^2)^2} + \frac{x}{16a^2(a^2+x^2)} + \frac{1}{16a^3}\arctan\left(\frac{x}{a}\right) \tag{C.76}$$

$$\int \frac{1}{a^4+x^4}\,\mathrm{d}x = \frac{\sqrt{2}}{8a^3}\left\{\ln\left(\frac{x^2+ax\sqrt{2}+a^2}{x^2-ax\sqrt{2}+a^2}\right) + 2\arctan\left(\frac{x\sqrt{2}}{a}-1\right) + 2\arctan\left(\frac{x\sqrt{2}}{a}+1\right)\right\} \tag{C.77}$$

$$\int \frac{x^2}{a^4+x^4}\,\mathrm{d}x = \frac{\sqrt{2}}{8a}\left\{-\ln\left(\frac{x^2+ax\sqrt{2}+a^2}{x^2-ax\sqrt{2}+a^2}\right) + 2\arctan\left(\frac{x\sqrt{2}}{a}-1\right) + 2\arctan\left(\frac{x\sqrt{2}}{a}+1\right)\right\} \tag{C.78}$$

C.3.4 Trigonometric Functions

$$\int \cos x\,\mathrm{d}x = \sin x \tag{C.79}$$

$$\int x\cos(ax)\,\mathrm{d}x = \frac{1}{a^2}[\cos(ax) + ax\sin(ax)] \tag{C.80}$$

$$\int x^2\cos(ax)\,\mathrm{d}x = \frac{1}{a^3}[2ax\cos(ax) + (a^2x^2-2)\sin(ax)] \tag{C.81}$$

$$\int \sin x\,\mathrm{d}x = -\cos x \tag{C.82}$$

$$\int x\sin(ax)\,\mathrm{d}x = \frac{1}{a^2}[\sin(ax) - ax\cos(ax)] \tag{C.83}$$

$$\int x^2\sin(ax)\,\mathrm{d}x = \frac{1}{a^3}[2ax\sin(ax) - (ax^2-2)\cos(ax)] \tag{C.84}$$

Reduction formulae:

$$\int \sin^n x\,\mathrm{d}x = \frac{-1}{n}\sin^{n-1} x\,\cos x + \frac{n-1}{n}\int \sin^{n-2} x\,\mathrm{d}x \tag{C.85}$$

$$\int \cos^n x\,\mathrm{d}x = \frac{1}{n}\cos^{n-1} x\,\sin x + \frac{n-1}{n}\int \cos^{n-2} x\,\mathrm{d}x \tag{C.86}$$

$$\int \tan^n x\,\mathrm{d}x = \frac{1}{n-1}\tan^{n-1} x - \int \tan^{n-2} x\,\mathrm{d}x, \quad n \neq 1 \tag{C.87}$$

C.3.5 Exponential Functions

$$\int \exp(ax)\,\mathrm{d}x = \frac{\exp(ax)}{a}, \quad a \text{ real or complex} \tag{C.88}$$

$$\int x\exp(ax)\,\mathrm{d}x = \exp(ax)\left(\frac{x}{a} - \frac{1}{a^2}\right), \quad a \text{ real or complex} \tag{C.89}$$

$$\int x^2\exp(ax)\,\mathrm{d}x = \exp(ax)\left(\frac{x^2}{a} - \frac{2x}{a^2} + \frac{2}{a^3}\right), \quad a \text{ real or complex} \tag{C.90}$$

$$\int x^3\exp(ax)\,\mathrm{d}x = \exp(ax)\left(\frac{x^3}{a} - \frac{3x^2}{a^2} + \frac{6x}{a^3} - \frac{6}{a^4}\right), \quad a \text{ real or complex} \tag{C.91}$$

$$\int x\exp(ax^2)\,\mathrm{d}x = \frac{1}{2a}\exp(ax^2) \tag{C.92}$$

$$\int \exp(ax)\sin(bx)\,\mathrm{d}x = \frac{\exp(ax)}{a^2+b^2}[a\sin(bx) - b\cos(bx)] \tag{C.93}$$

$$\int \exp(ax)\cos(bx)\,\mathrm{d}x = \frac{\exp(ax)}{a^2+b^2}[a\cos(bx) + b\sin(bx)] \tag{C.94}$$

Reduction formula:

$$\int \frac{\exp x}{x^n}\,\mathrm{d}x = -(n-1)\frac{\exp x}{x^{n-1}} + (n-1)\int \frac{\exp x}{x^{n-1}}\,\mathrm{d}x \tag{C.95}$$

C.4 DEFINITE INTEGRALS

$$\int_{-\infty}^{\infty} \exp(-ax^2)\,\mathrm{d}x = \sqrt{\frac{\pi}{a}}, \quad a > 0 \tag{C.96}$$

$$\int_{-\infty}^{\infty} \exp(-a^2x^2 + bx)\,\mathrm{d}x = \frac{\sqrt{\pi}}{a}\exp\left(\frac{b^2}{4a^2}\right), \quad a > 0 \tag{C.97}$$

$$\int_{0}^{\infty} x^2\exp(-ax^2)\,\mathrm{d}x = \frac{1}{4a}\sqrt{\frac{\pi}{a}}, \quad a > 0 \tag{C.98}$$

$$\int_0^\infty \frac{\sin x}{x}\,\mathrm{d}x = \frac{\pi}{2} \tag{C.99}$$

$$\int_0^\infty \left(\frac{\sin x}{x}\right)^2 \mathrm{d}x = \frac{\pi}{2} \tag{C.100}$$

$$\int_0^\infty \frac{x\sin(ax)}{b^2+x^2}\,\mathrm{d}x = \frac{\pi}{2}\exp(-ab), \quad a>0, b>0 \tag{C.101}$$

$$\int_0^\infty \frac{\cos(ax)}{b^2+x^2}\,\mathrm{d}x = \frac{\pi}{2b}\exp(-ab), \quad a>0, b>0 \tag{C.102}$$

$$\int_0^\infty \frac{\cos(ax)}{(b^2-x^2)^2}\,\mathrm{d}x = \frac{\pi}{4b^3}[\sin(ab) - ab\cos(ab)], \quad a>0, b>0 \tag{C.103}$$

C.5 SERIES

Taylor expansion:

$$f(x+a) = f(x) + af'(x) + \frac{a^2}{2!}f''(x) + \cdots + \frac{a^n}{n!}f^{(n)}(x) + \cdots \tag{C.104}$$

McLaurin expansion:

$$f(x) = f(0) + xf'(0) + \frac{x^2}{2!}f''(0) + \cdots + \frac{x^n}{n!}f^{(n)}(0) + \cdots \tag{C.105}$$

$$\exp x = 1 + x + \frac{1}{2!}x^2 + \frac{1}{3!}x^3 + \cdots \tag{C.106}$$

$$\sin x = x - \frac{1}{3!}x^3 + \frac{1}{5!}x^5 - \cdots \tag{C.107}$$

$$\cos x = 1 - \frac{1}{2!}x^2 + \frac{1}{4!}x^4 - \cdots \tag{C.108}$$

$$\tan x = x + \frac{1}{3}x^3 + \frac{2}{15}x^5 + \cdots \tag{C.109}$$

$$\arcsin x = x + \frac{1}{6}x^3 + \frac{3}{40}x^5 + \cdots \tag{C.110}$$

$$\arctan x = x - \frac{1}{3}x^3 + \frac{1}{5}x^5 - \cdots, \quad |x| < 1 \tag{C.111}$$

$$\operatorname{sinc} x = 1 - \frac{1}{3!}x^2 + \frac{1}{5!}x^4 - \cdots \tag{C.112}$$

$$\ln(1+x) = x - \frac{1}{2}x^2 + \frac{1}{3}x^3 - \cdots \tag{C.113}$$

$$(1+x)^n = 1 + nx + \frac{n(n-1)}{2!}x^2 + \frac{n(n-1)(n-2)}{3!}x^3 + \cdots, \quad |nx| < 1 \tag{C.114}$$

$$\sum_{n=1}^{N} n = \frac{N(N+1)}{2} \tag{C.115}$$

$$\sum_{n=1}^{N} n^2 = \frac{N(N+1)(2N+1)}{6} \tag{C.116}$$

$$\sum_{n=1}^{N} n^3 = \frac{N^2(N+1)^2}{4} \tag{C.117}$$

$$\sum_{n=0}^{N} x^n = \frac{1-x^{N+1}}{1-x} \tag{C.118}$$

$$\sum_{n=0}^{N} \frac{N!}{n!(N-n)!} x^n y^{N-n} = (x+y)^N \tag{C.119}$$

$$\sum_{n=0}^{\infty} ax^n = \frac{a}{1-x}, \qquad |x| < 1 \tag{C.120}$$

C.6 LOGARITHMS

Definitions:

$$\text{If } 10^x = y, \quad \text{then} \quad \log y \triangleq x, \quad y \geq 0 \tag{C.121}$$

$$\text{If } \exp\ x = y, \quad \text{then} \quad \ln y \triangleq x, \quad y \geq 0 \tag{C.122}$$

Properties:

$$\log(ab) = \log a + \log b \tag{C.123}$$

$$\log\left(\frac{a}{b}\right) = \log a - \log b \tag{C.124}$$

$$\ln(ab) = \ln a + \ln b \tag{C.125}$$

$$\ln\left(\frac{a}{b}\right) = \ln a - \ln b \tag{C.126}$$

$$\log a^b = b \log a \tag{C.127}$$

$$\ln a^b = b \ln a \tag{C.128}$$

$$\log a = \frac{\ln a}{\ln(10)} = 0.4343 \times \ln a \tag{C.129}$$

Appendix D

Summary of Probability Theory

1. *Probability distribution function:*

$$F_X(x) = \mathrm{P}(X \leq x) = \int_{-\infty}^{x} f_X(u)\,\mathrm{d}u$$

2. *Probability density function:*

$$f_X(x) = \frac{\mathrm{d}F_X(x)}{\mathrm{d}x}$$

3. *Gaussian probability density function:*

$$f_X(x) = \frac{1}{\sigma\sqrt{2\pi}}\exp\left[-\frac{(x-\overline{X})^2}{2\sigma^2}\right]$$

4. *Independence:*

$$f_{X,Y}(x,y) = f_X(x)\,f_Y(y)$$

5. *Expectation:*

$$\mathrm{E}[X] = \int_{-\infty}^{\infty} x f_X(x)\,\mathrm{d}x$$

$$\mathrm{E}[g(X)] = \int_{-\infty}^{\infty} g(x) f_X(x)\,\mathrm{d}x \rightarrow \text{special case:}\quad \mathrm{E}[X^2] = \int_{-\infty}^{\infty} x^2 f_X(x)\,\mathrm{d}x$$

Introduction to Random Signals and Noise W. van Etten

6. *Variance:*

$$\sigma_X^2 \triangleq \mathrm{E}[(X - \mathrm{E}[X])^2] = \mathrm{E}[X^2] - \mathrm{E}^2[X]$$

7. *Addition of random variables:*

$$\mathrm{E}[X + Y] = \mathrm{E}[X] + \mathrm{E}[Y]$$

8. *Independent random variables:*

$$\mathrm{E}[X\,Y] = \mathrm{E}[X]\ \mathrm{E}[Y]$$

9. *Correlation:*

$$R_{XY} = \mathrm{E}[XY] = \iint_{-\infty}^{\infty} xy\, f_{X,Y}(x, y)\, \mathrm{d}x\, \mathrm{d}y$$

Appendix E
Definition of a Few Special Functions

1. *Unit-step function (Figure E.1):*

$$\mathrm{u}(x) \triangleq \begin{cases} 1, & x \geq 0 \\ 0, & x < 0 \end{cases}$$

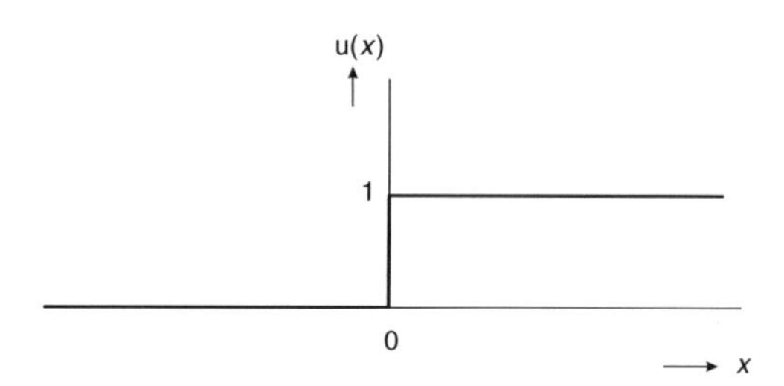

Figure E.1

2. *Signum function (Figure E.2):*

$$\mathrm{sgn}(x) \triangleq \begin{cases} 1, & x \geq 0 \\ -1, & x < 0 \end{cases}$$

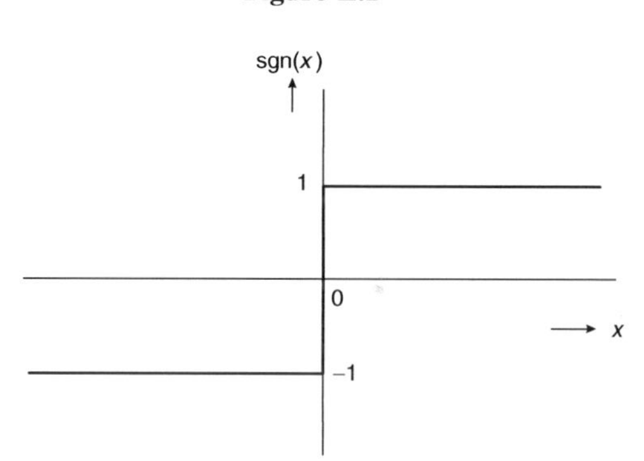

Figure E.2

3. *Rectangular function (Figure E.3):*

$$\mathrm{rect}(x) \triangleq \begin{cases} 1, & |x| \leq \frac{1}{2} \\ 0, & |x| > \frac{1}{2} \end{cases}$$

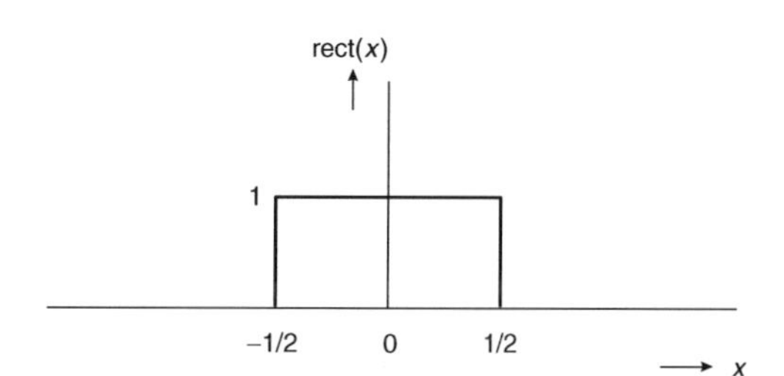

Figure E.3

Introduction to Random Signals and Noise W. van Etten

4. *Triangular function (Figure E.4):*

$$\mathrm{tri}(x) \triangleq \begin{cases} 1 - |x|, & |x| \le 1 \\ 0, & |x| > 1 \end{cases}$$

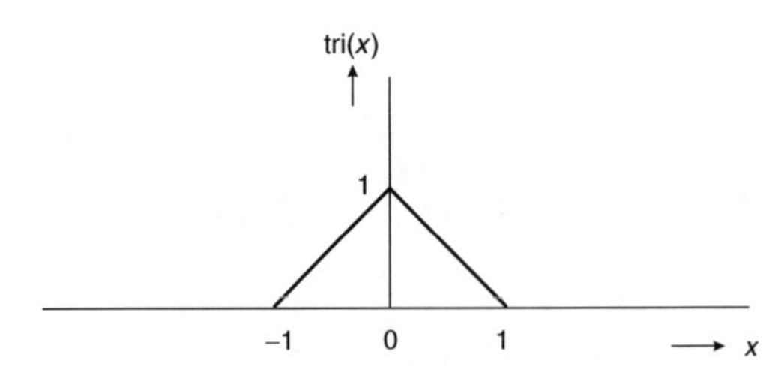

Figure E.4

5. *Sinc function (Figure E.5):*

$$\mathrm{sinc}(x) \triangleq \frac{\sin x}{x}$$

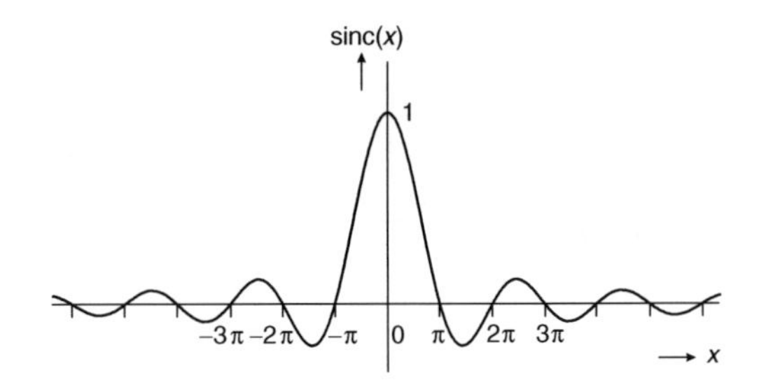

Figure E.5

6. *Delta function (Figure E.6):*

$$\int_a^b f(x)\delta(x - x_0)\,\mathrm{d}x \triangleq \begin{cases} f(x_0), & a \le x_0 < b \\ 0, & \text{elsewhere} \end{cases}$$

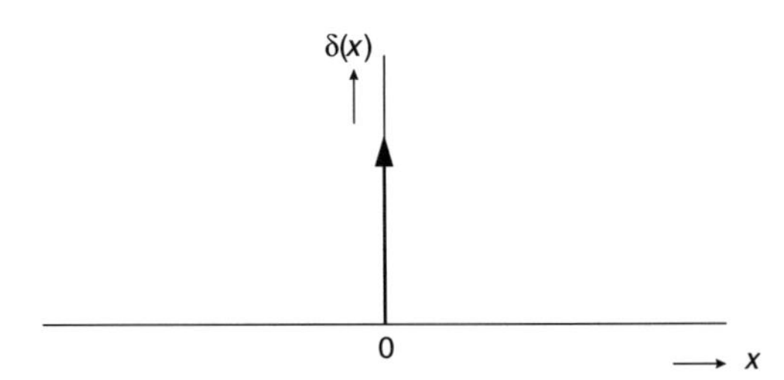

Figure E.6

Appendix F
The Q(.) and erfc Functions

The Q function is defined as

$$\mathrm{Q}(x) \triangleq \frac{1}{\sqrt{2\pi}} \int_x^{\infty} \exp\left(-\frac{y^2}{2}\right) \mathrm{d}y \tag{F.1}$$

The function is used to evaluate the error probability of transmission systems that are disturbed by additive Gaussian noise. Some textbooks use a different function for that purpose, namely the complementary error function, abbreviated as erfc. This latter function is defined as

$$\mathrm{erfc}(x) \triangleq 1 - \mathrm{erf}(x) = \frac{2}{\sqrt{\pi}} \int_x^{\infty} \exp(-y^2)\, \mathrm{d}y \tag{F.2}$$

From Equations (F.1) and (F.2) it follows that the Q function is related to the erfc function as follows:

$$\mathrm{Q}(x) = \frac{1}{2} \mathrm{erfc}\left(\frac{x}{\sqrt{2}}\right) \tag{F.3}$$

The integral in these equations cannot be solved analytically. A simple and accurate expression (error less than 0.27 %) is given by

$$\mathrm{Q}(x) \approx \left[\frac{1}{(1-0.339)x + 0.339\sqrt{x^2+5.510}}\right] \frac{\exp(-x^2/2)}{\sqrt{2\pi}} \tag{F.4}$$

Most modern mathematical software packages such as Matlab, Maple and Mathematica comprise the erfc function as a standard function. Both functions are presented graphically in Figures F.1 and F.2.

Introduction to Random Signals and Noise W. van Etten

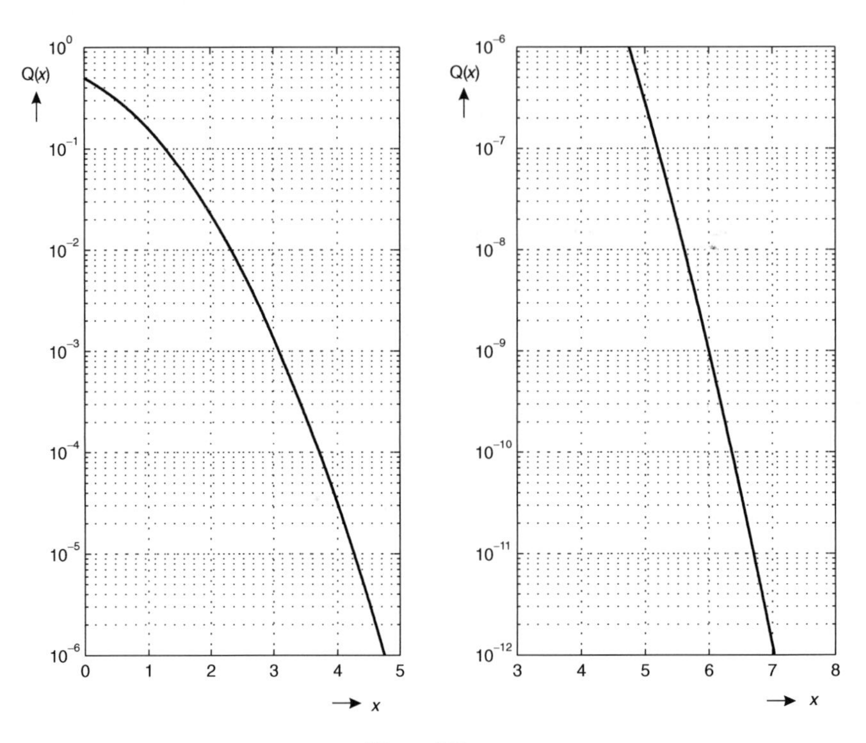

Figure F.1

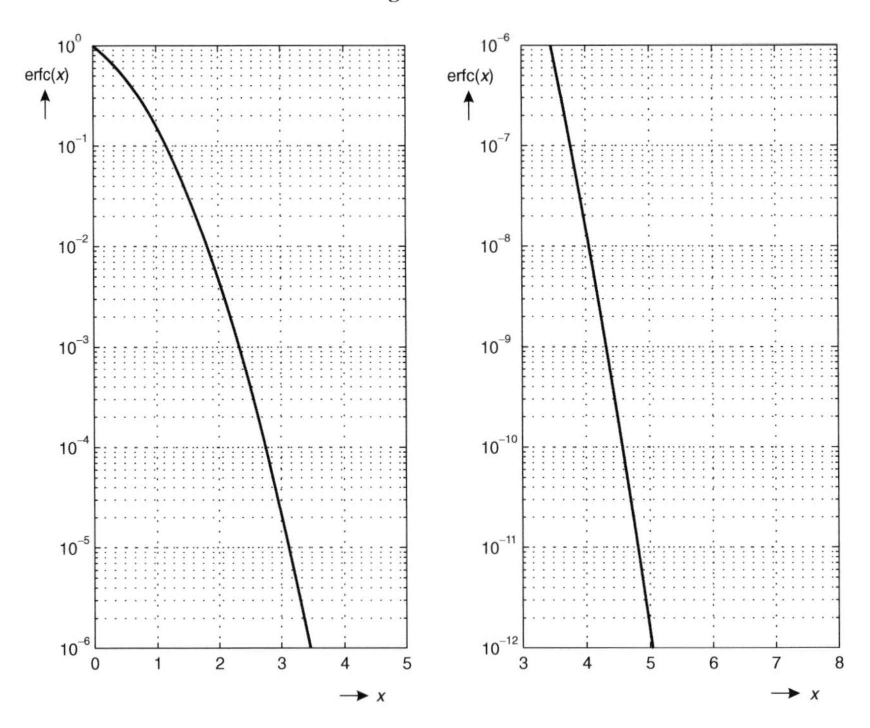

Figure F.2

Appendix G
Fourier Transforms

Definition:

$$X(\omega) = \int_{-\infty}^{\infty} x(t)\exp(-\mathrm{j}\omega t)\,\mathrm{d}t \Longleftrightarrow x(t) = \frac{1}{2\pi}\int_{-\infty}^{\infty} X(\omega)\exp(\mathrm{j}\omega t)\,\mathrm{d}\omega$$

Properties:

Time domain	Frequency domain
1. $ax_1(t) + bx_2(t)$	$aX_1(\omega) + bX_2(\omega)$
2. $x(at)$	$\frac{1}{\lvert a\rvert}X(\frac{\omega}{a})$
3. $x(-t)$	$X^*(\omega)$
4. $x(t - t_0)$	$X(\omega)\exp(-\mathrm{j}\omega t_0)$
5. $x(t)\exp(\mathrm{j}\omega_0 t)$	$X(\omega - \omega_0)$
6. $\frac{\mathrm{d}^n x(t)}{\mathrm{d}t^n}$	$(\mathrm{j}\omega)^n X(\omega)$
7. $\int_{-\infty}^{t} x(\tau)\,\mathrm{d}\tau$	$\frac{X(\omega)}{\mathrm{j}\omega} + \pi X(0)\delta(\omega)$
8. $\int_{-\infty}^{\infty} x(t)\,\mathrm{d}t \;=\rightarrow$	$\leftarrow= X(0)$
9. $x(0) \;=\rightarrow$	$\leftarrow= \frac{1}{2\pi}\int_{-\infty}^{\infty} X(\omega)\,\mathrm{d}\omega$
10. $(-\mathrm{j}t)^n x(t)$	$\frac{\mathrm{d}^n X(\omega)}{\mathrm{d}\omega^n}$
11. $x^*(t)$	$X^*(-\omega)$
12. $\int_{-\infty}^{\infty} x_1(\tau)x_2(t-\tau)\,\mathrm{d}\tau$(conv.)	$X_1(\omega)X_2(\omega)$
13. $x_1(t)x_2(t)$	$\frac{1}{2\pi}\int_{-\infty}^{\infty} X_1(\rho)X_2(\omega-\rho)\,\mathrm{d}\rho$ (conv.)
14. $\int_{-\infty}^{\infty} \lvert x(t)\rvert^2\,\mathrm{d}t \;=\rightarrow$	$\leftarrow= \frac{1}{2\pi}\int_{-\infty}^{\infty} \lvert X(\omega)\rvert^2\,\mathrm{d}\omega$ (Parseval)
15. $X(t)$	$2\pi x(-\omega)$(duality)

Fourier table with α, τ, σ, ω_0 and W real constants:

$x(t)$	$X(\omega)$	Condition
1. $\alpha\delta(t)$	α	
2. $\dfrac{\alpha}{2\pi}$	$\alpha\delta(\omega)$	
3. $u(t)$	$\pi\delta(\omega)+\frac{1}{j\omega}$	
4. $\frac{1}{2}\delta(t)-\dfrac{1}{j2\pi t}$	$u(\omega)$	
5. $\mathrm{rect}(t/\tau)$	$\tau\,\mathrm{sinc}(\omega\tau/2)$	$\tau>0$
6. $(W/\pi)\,\mathrm{sinc}(Wt)$	$\mathrm{rect}\left(\dfrac{\omega}{2W}\right)$	$W>0$
7. $\mathrm{tri}(t/\tau)$	$\tau\,\mathrm{sinc}^2(\omega\tau/2)$	$\tau>0$
8. $(W/\pi)\mathrm{sinc}^2(Wt)$	$\mathrm{tri}\left(\dfrac{\omega}{2W}\right)$	$W>0$
9. $\mathrm{sgn}(t)$	$\dfrac{2}{j\omega}$	
10. $\dfrac{-1}{j\pi t}$	$\mathrm{sgn}(\omega)$	
11. $\exp(j\omega_0 t)$	$2\pi\delta(\omega-\omega_0)$	
12. $\delta(t-\tau)$	$\exp(-j\omega\tau)$	
13. $\cos(\omega_0 t)$	$\pi[\delta(\omega-\omega_0)+\delta(\omega+\omega_0)]$	
14. $\sin(\omega_0 t)$	$-j\pi[\delta(\omega-\omega_0)-\delta(\omega+\omega_0)]$	
15. $u(t)\cos(\omega_0 t)$	$\dfrac{\pi}{2}[\delta(\omega-\omega_0)+\delta(\omega+\omega_0)]+\dfrac{j\omega}{\omega_0^2-\omega^2}$	
16. $u(t)\sin(\omega_0 t)$	$-j\dfrac{\pi}{2}[\delta(\omega-\omega_0)-\delta(\omega+\omega_0)]+\dfrac{\omega_0}{\omega_0^2-\omega^2}$	
17. $u(t)\exp(-\alpha t)$	$\dfrac{1}{\alpha+j\omega}$	$\alpha>0$
18. $u(t)t\exp(-\alpha t)$	$\dfrac{1}{(\alpha+j\omega)^2}$	$\alpha>0$
19. $u(t)t^2\exp(-\alpha t)$	$\dfrac{2}{(\alpha+j\omega)^3}$	$\alpha>0$
20. $u(t)t^3\exp(-\alpha t)$	$\dfrac{6}{(\alpha+j\omega)^4}$	$\alpha>0$
21. $\exp(-\alpha\lvert t\rvert)$	$\dfrac{2\alpha}{\alpha^2+\omega^2}$	$\alpha>0$
22. $\dfrac{1}{\sigma\sqrt{2\pi}}\exp\left(\dfrac{-t^2}{2\sigma^2}\right)$	$\exp\left(\dfrac{-\sigma^2\omega^2}{2}\right)$	$\sigma>0$
23. $\sum_{n=-\infty}^{\infty}\delta(t-nT)$	$\dfrac{2\pi}{T}\sum_{n=-\infty}^{\infty}\delta\left(\omega-n\dfrac{2\pi}{T}\right)$	

Appendix H
Mathematical and Physical Constants

Base of natural logarithm:	$e = 2.718\,281\,8$
Logarithm of 2 to base 10:	$\log(2) = 0.301\,030\,0$
Pi:	$\pi = 3.141\,592\,7$
Boltzmann's constant:	$k = 1.38 \times 10^{-23}$ [J/K]
Planck's constant:	$h = 6.63 \times 10^{-34}$ [J s]
Temperature in kelvin:	Temperature in °C + 273
Standard ambient temperature:	$T_0 = 290$ [K] = 17 [°C]
Thermal energy kT at standard ambient temperature:	$kT_0 = 4.00 \times 10^{-21}$ [J]

Introduction to Random Signals and Noise W. van Etten

Index

Introduction to Random Signals and Noise W. van Etten

Printed and bound in the UK by
CPI Antony Rowe, Eastbourne

Printed and bound by CPI Group (UK) Ltd, Croydon, CR0 4YY
17/04/2025
14658868-0001